Longman Mathematical Texts

The solution of ordinary differential equations

Longman Mathematical Texts

The solution of ordinary differential equations

The late **E.L. Ince**
and **I.N. Sneddon** F.R.S.
Emeritus Professor of Mathematics
in the University of Glasgow

Copublished in the United States with
John Wiley & Sons, Inc., New York

Longman Scientific & Technical,
Longman Group UK Limited
Longman House, Burnt Mill, Harlow
Essex CM20 2JE, England
and Associated companies throughout the world

Copublished in the United States with
John Wiley & Sons, Inc., 605 *Third Avenue, New York, NY* 10158

This edition published 1987

This is a new edition based on the seventh edition (1963 reprint) of *Integration of Ordinary Differential Equations*, first published by Oliver & Boyd, 1939.

British Library Cataloguing in Publication Data

Ince, E.L.
The solution of ordinary differential equations. ——— (Longman mathematical texts)
1. Differential equations
I. Title II. Sneddon, I.N.
515.3′52

ISBN 0-582-44068-8

Library of Congress Cataloging-in-Publication Data

Ince, Edward Lindsay, 1891–1941.
The solution of ordinary differential equations.

(Longman mathematical texts)
Rev. ed. of: Integration of ordinary differential equations. 7th ed. 1963.
Includes index.
1. Differential equations. I. Sneddon, Ian Naismith.
II. Ince, Edward Lindsay, 1891–1941. Integration of ordinary differential equations. III. Title. IV. Series.
QA372.I6 1987 515.3′52 86-2945
ISBN 0-470-20680-2 (USA only)

Set in Monophoto Times New Roman
Produced by Longman Group (FE) Limited
Printed in Hong Kong

Contents

4: Equation of the second and higher orders

5: Linear equations

6: Solution in series

7: Linear systems

Appendices

Preface

In the Preface to the First Edition of the University Mathematical Text, *Integration of Ordinary Differential Equations* (Oliver & Boyd, Edinburgh, 1939), the late Dr. E.L. Ince wrote:

'The object of this book is to provide in a compact form an account of the methods of integrating explicitly the commoner types of ordinary differential equation, and in particular those equations that arise from problems in geometry and applied mathematics. It takes the existence of solutions for granted; the reader who desires to look into the theoretical background of the methods here outlined will find what he seeks in my larger treatise *Ordinary Differential Equations* (Longmans, Green & Co. Ltd., 1927). With this qualification, it will be found to contain all the material needed by students of mathematics in our Universities who do not specialize in differential equations, as well as by students of mathematical physics and technology.

As one of the first things a beginner has to learn is to identify the type to which a given equation belongs, the problems for solution have not been printed after the sections to which they refer, but have been collected at the end of the book. When the contents of the first chapter have been mastered, the reader may test his skill by attacking examples selected at random from Nos. 1 to 122, and similarly for the later chapters. At the same time, the examples occur roughly in the order of the table of contents, so that working material is always available as reading progresses.'

Dr. Ince died in 1941 at the age of 50, having returned to Edinburgh in 1932 as Lecturer in Technical Mathematics in the University of Edinburgh, after a period of six years as Professor of Pure Mathematics in The Egyptian University. Because of his untimely death, the Second Edition of the smaller book had not the advantage of being revised by him, but a scrutiny of the book and revision of parts of it were undertaken by the late Professor A. Erdélyi, F.R.S. The popularity of the book is evidenced by the fact that by 1956 it had gone to seven editions. The present book is based on the Second Reprint (1963) of that Seventh Edition. In the intervening years the demand for a book of the type of

Ince's little book has been undiminished. When I was approached by a publisher to write a book just slightly larger, I suggested that it would be far better to use the material in Ince's book and merely to bring it up-to-date with modern practice and to add a further seventy-five problems to test the reader's comprehension of the added material. As far as possible I have sought to preserve the content and the structure of the original text. I have confined by own contributions to the inclusion of a few sections and problems wherever these seemed to be needed. Above all I hope I have been successful in preserving Ince's original intention: to provide in a compact form an introduction to the methods of solution of ordinary differential equations for students of mathematics taking a first course in the study of differential equations and their solution, and for students in science and technology who, for their work in these subjects, have to acquire a feel for the nature of the solutions to differential equations before going on to study their numerical solution in the computing laboratory.

1

Equations of the first order and degree

§1 Definitions

We suppose that $y(x)$ is a function of the independent real variable x, and denote by $y', y'', \ldots, y^{(n)}$ successive derivatives of y with respect to x. Any relation of equality which involves at least one of these derivatives is said to be an *ordinary differential equation.* The term *ordinary* distinguishes such an equation from a partial differential equation, which involves two or more independent variables, a function of these variables and the corresponding partial derivatives. The order of any differential equation is the order of the highest derivative involved. Thus any relation of the form

$$F(x, y, y', y'', \ldots, y^{(n)}) = 0,$$

is an ordinary differential equation of order n. Normally x is a distance-like variable but in applications we frequently take t, the time, to be the independent variable and write $\dot{x}$ for the derivative with respect to t of $x(t)$.

(a) The geometry of plane curves

The simplest illustration of the occurrence of ordinary differential equations may be taken from the geometry of plane curves.

$$f(x, y, c) = 0, \tag{1.1}$$

in which x and y are rectangular coordinates and c is a parameter, represents a family of plane curves, in which a specific value of the parameter c corresponds to a unique curve of the family. If, regarding c for the moment as fixed, we differentiate with respect to x, we obtain the equation

$$\frac{\partial f}{\partial x} + \frac{\partial f}{\partial y} y' = 0. \tag{1.2}$$

In general, the latter equation involves c; if c is eliminated between equations (1.1) and (1.2), the result is an equation involving x, y and y',

$$F(x, y, y')=0, \tag{1.3}$$

that is, an ordinary differential equation *of the first order*. When such an equation is a polynomial in y' (but not necessarily in x and y) the index of the highest power of y' is said to be the *degree* of the equation.

Geometrically, the differential equation (1.3) implies that at any chosen point of the xy-plane the derivative has a certain value or values, that is to say it symbolizes a property of the gradient of any curve of the value (1.1) that passes through the point (x, y) under consideration.

Example 1.1 The equation

$$y=x^2+c$$

represents a family of equal parabolas having the y-axis as their common axis. On differentiating with respect to x we have the first-order differential equation

$$y'=2x.$$

The arbitrary constant c has disappeared, so that this is actually the differential equation of the family of parabolas. It expresses the fact that all the cuves of the family have the same gradient at the points where they are cut by a line parallel to the y-axis, namely a gradient equal to twice the abscissa of the line.

Example 1.2 The equation

$$y=cx^2$$

represents a family of similar parabolas having the y-axis as their common axis, and all touching the x-axis at the origin. Differentiating, we obtain

$$y'=2cx,$$

which involves c; if this constant is eliminated we obtain the differential equation

$$xy'=2y$$

which expresses the property that any line with an equation of the form $y=mx$ through the origin intersects all curves of the family in points where they have the same gradient $2m$.

The process of elimination by which the differential equation (1.3) was obtained from the primitive equation (1.1) is in general not reversible; the action of recovering the primitive, or an equivalent expression, is known as integration. More precisely, to integrate or solve a differential equation of the first order is to determine all the relations $f(x, y)=0$ such that the values of y and y' deduced from them in terms of x shall satisfy the differential equation identically.

When an infinite set of such integrals can be grouped in one comprehensive formula, involving an arbitrary constant c, say

$$f(x, y, c)=0,$$

it is known as a general integral; it is in fact either the primitive or an expression equivalent to it. If two integrals exist, each of which involves an arbitrary constant, they can be transformed into one another; see Example 3.2 below for an example of this situation. Any integral that can be obtained by assigning a definite numerical value to the constant c is called a *particular* integral. However, there may be integrals other than those that can be obtained by assigning particular values to c; these are called *singular* integrals.

Example 1.3 The equation

$$(y')^2 - xy' + y=0$$

(which is of the first order and second degree) possesses the general integral

$$y=cx-c^2.$$

This represents a family of straight lines, and any particular integral corresponds to a definite line in the family (cf. Fig. 1). However the equation is also satisfied by

$$y=\tfrac{1}{4}x^2,$$

which represents not a straight line but a parabola. This is a singular integral.

(b) Motion of a particle in a straight line

We now consider the motion of a particle of mass m whose distance at time t from a fixed origin is $x(t)$; the velocity of the particle is $v(t)=\dot{x}(t)$ and its acceleration is $f(t)=\dot{v}(t)=\ddot{x}(t)$, where $\ddot{x}(t)$ denotes the second derivative of $x(t)$ with respect to t. If the force applied to the particle to produce the motion is $F(x, \dot{x}, t)$ then, by Newton's second law the motion

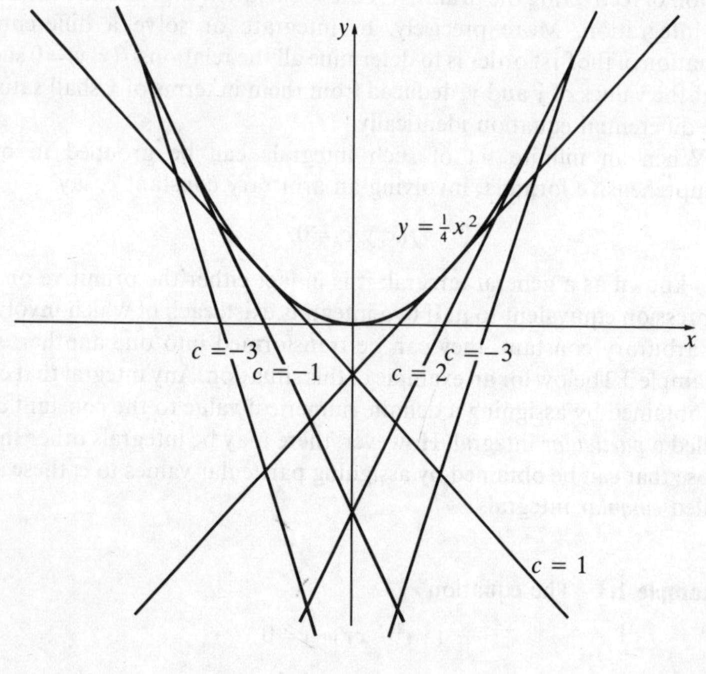

Fig. 1

is described by the differential equation

$$m\ddot{x}(t) = F(x, \dot{x}, t) \tag{1.4}$$

and if the particle starts off at the time $t=0$ at the point $x=a$ with prescribed velocity U the initial conditions are

$$x(0) = a, \qquad \dot{x}(0) = U. \tag{1.5}$$

This is, of course, an ordinary differential equation of the *second* order but for certain forms of the function F the motion may be described by a first-order equation. For instance, if $F = F_1(\dot{x}, t)$ we may rewrite equation (1.4) as

$$m\dot{v} = F_1(v, t) \tag{1.6}$$

where $v = \dot{x}$, and the initial conditions (1.5) as the single equation

$$mv(0) = U. \tag{1.7}$$

The problem of determining the solution of equation (1.6) which satisfies the condition (1.7) is called the initial-value problem for the first-order equation (1.6). If the solution of this equation is denoted by $v(t; U)$, then since $x = v$ and $x(0) = a$ we find that the distance travelled by the particle in time t is

$$x(t) = a + \int_0^t v(s; U)\, \mathrm{d}s \tag{1.8}$$

Similarly, if $F = F_2(x, v)$ we can use the formula

$$v(t) = v'(x)\dot{x}(t) = vv'(x)$$

to rewrite equation (1.4) in the form

$$v'(x) = F_3(x, v) \tag{1.9}$$

with $F_3(x, v) = v^{-1}F_2(x, v)$, and the boundary conditions (1.5) in the alternative form

$$v(a) = U; \tag{1.10}$$

the basic problem therefore reduces to finding the solution $v(x; a, U)$ of the initial-value problem posed by equations (1.9) and (1.10). Once this solution has been found we may use the formula

$$\frac{\mathrm{d}t}{\mathrm{d}x} = \frac{1}{v}$$

to derive the relation

$$t = \int_a^x \frac{\mathrm{d}\xi}{v(\xi; a, U)}.$$

In the general case we may reduce the solution of the initial-value problem posed by equations (1.4) and (1.5) to that of a pair of first-order differential equations. If we introduce the momentum $p = m\dot{x}$, we can write equation (1.4) as the pair

$$\dot{x} = p/m, \qquad \dot{p} = F(x, p/m, t) \tag{1.11}$$

and equations (1.6) as

$$x(0) = a, \qquad p(0) = U/m. \tag{1.12}$$

Introducing the 2-vectors

$$\mathbf{X}(t) = \begin{bmatrix} x \\ p \end{bmatrix} \qquad \mathbf{G}(\mathbf{X}, t) = \begin{bmatrix} p/m \\ F(x, p/m, t) \end{bmatrix} \qquad \mathbf{X}^0 = \begin{bmatrix} a \\ U/m \end{bmatrix}$$

we may rewrite these equations as the pair of equations

$$\dot{\mathbf{X}}(t) = \mathbf{G}(\mathbf{X}, t), \qquad \mathbf{X}(0) = \mathbf{X}^0. \tag{1.13}$$

We can therefore regard the solution of the second-order initial-value problem (1.4), (1.5) as equivalent to solving the pair of simultaneous first-order differential equations (1.11) or, alternatively, the vector equation, again of the first order, (1.13).

(c) Motion of a particle in the plane

We can similarly describe the motion of a particle in a plane in terms of a pair of rectangular coordinates (x_1, x_2). The equations of motion of a particle of mass m are of the form

$$m\ddot{x}_1 = F_1(x_1, x_2, \dot{x}_1, \dot{x}_2; t),$$
$$m\ddot{x}_2 = F_2(x_1, x_2, \dot{x}_1, \dot{x}_2; t)$$

and the initial conditions are of the form

$$x_1(0) = a^0, \quad x_2(0) = b^0; \quad \dot{x}_1(0) = u^0, \quad \dot{x}_2(0) = v^0.$$

By introducing the momenta $p_1 = m\dot{x}_1(t)$, $p_2 = m\dot{x}_2(t)$ we can rewrite the pair of second-order differential equations by the four first-order equations

$$\dot{x}_1 = p_1/m \quad \dot{x}_2 = p_2/m, \quad \dot{p}_1 = F_1(x_1, x_2, p_1/m, p_2/m; t),$$
$$\dot{p}_2 = F_2(x_1, x_2, p_1/m, p_2/m; t)$$

and the initial conditions by

$$x_1(0) = a^0, \quad x_2(0) = b^0; \quad p_1(0) = u^0/m, \quad p_2(0) = v^0/m.$$

Also by introducing the vectors

$$\mathbf{X}(t) = \begin{bmatrix} x_1 \\ x_2 \\ p_1 \\ p_2 \end{bmatrix}, \qquad \mathbf{G}(\mathbf{X}, t) = \begin{bmatrix} p_1/m \\ p_2/m \\ F_1 \\ F_2 \end{bmatrix}, \qquad \mathbf{X}^0 = \begin{bmatrix} a^0 \\ b^0 \\ u^0/m \\ v^0/m \end{bmatrix}$$

we can reduce the pair of second-order differential equations to the initial-value problem of the first order (1.13).

For example the motion of a particle moving under gravity against a resistance proportional to its velocity is described by the pair of second-

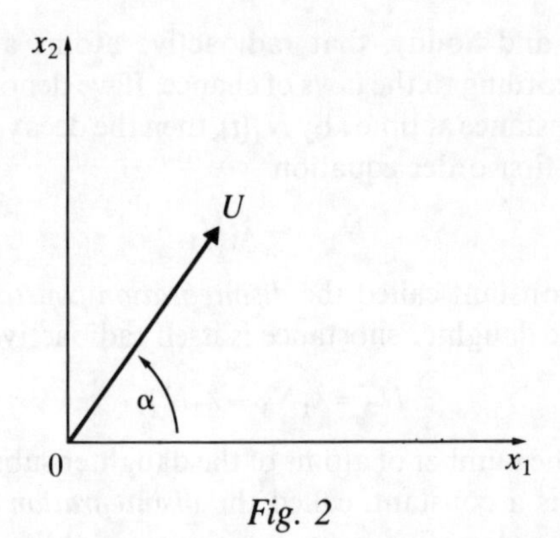

Fig. 2

order equations

$$m\ddot{x}_1 = -k\dot{x}_1, \qquad m\ddot{x}_2 = -k\dot{x}_2 - g,$$

where k and g are positive constants, and the initial conditions

$$x_1(0)=0, \quad x_2(0)=0, \quad \dot{x}_1(0)=U\cos\alpha, \quad \dot{x}_2(0)=U\sin\alpha$$

(cf. Fig. 2). In terms of the momenta p_1 and p_2, the above vectors become

$$\mathbf{G}(\mathbf{X}, t) = \begin{bmatrix} p_1/m \\ p_2/m \\ -kp_1/m \\ -kp_2/m - g \end{bmatrix}, \quad \mathbf{X}^0 = \begin{bmatrix} 0 \\ 0 \\ U\cos\alpha/m \\ U\sin\alpha/m \end{bmatrix}.$$

We can write equation $\dot{\mathbf{X}}(t) = \mathbf{G}(\mathbf{X}, t)$ with the above form for $\mathbf{G}$ in the alternative form

$$\dot{\mathbf{X}} = A\mathbf{X} + \mathbf{b} \tag{1.14}$$

where

$$A = \begin{bmatrix} 0 & 0 & m^{-1} \\ 0 & 0 & 0 \\ 0 & 0 & -k/m \\ 0 & 0 & 0 \end{bmatrix}, \qquad \mathbf{b} = \begin{bmatrix} 0 \\ 0 \\ 0 \\ -g \end{bmatrix}. \tag{1.15}$$

(d) Theory of radioactive transformations

Another example of the occurrence of simultaneous first-order ordinary differential equations is provided by the theory, first put forward in 1902 by Rutherford and Soddy, that radioactive atoms are unstable and disintegrate according to the laws of chance. If we denote the number of atoms of the substance at time t by $N_1(t)$, then the decay of this substance is given by the first-order equation

$$\dot{N}_1 = -\lambda_1 N_1, \tag{1.16}$$

where λ_1 is a constant called the *disintegration constant* of the parent substance. If the daughter substance is itself radioactive, then, similarly

$$\dot{N}_2 = \lambda_1 N_1 - \lambda_2 N_2,$$

where $N_2(t)$ is the number of atoms of the daughter substance present at time t, and λ_2 is a constant, called the *disintegration constant* of that substance. The radioactive series continues in this way, the $(r-1)$th product decaying into the rth according to the rule

$$\dot{N}_{r+1} = \lambda_r N_r - \lambda_{r+1} N_{r+1}, \qquad 1 \leqslant r \leqslant n-1, \tag{1.17}$$

where if we assume that the nth product is stable it will increase according to the rule

$$\dot{N}_{n+1} = \lambda_n N_n. \tag{1.18}$$

Assuming that initially only the parent substance is present, we have the initial conditions

$$N_1(0) = N_0; \qquad N_2(0) = N_3(0) = \cdots = N_{n+1}(0) = 0 \tag{1.19}$$

Introducing the vector $\mathbf{N}$ and the matrix A through the equations

$$\mathbf{N} = \begin{bmatrix} N_1 \\ N_2 \\ N_3 \\ \vdots \\ N_n \\ N_{n+1} \end{bmatrix} \qquad A = \begin{bmatrix} -\lambda_1 & 0 & 0 & \dots & 0 & 0 \\ \lambda_1 & -\lambda_2 & 0 & \dots & 0 & 0 \\ 0 & \lambda_2 & -\lambda_3 & \dots & 0 & 0 \\ \vdots & \vdots & \vdots & & \vdots & \vdots \\ 0 & 0 & 0 & \dots & \lambda_{n-1} & -\lambda_n \\ 0 & 0 & 0 & \dots & 0 & \lambda_n \end{bmatrix}$$

we can write the system of first-order (1.16)–(1.8) by the first-order vector differential equation

$$\dot{\mathbf{N}} = A\mathbf{N} \tag{1.20}$$

and the initial condition (1.19) by the equation

$$\mathbf{N}(0) = \mathbf{N}^0, \tag{1.21}$$

where

$$\mathbf{N}^0 = \begin{bmatrix} N_0 \\ 0 \\ 0 \\ \vdots \\ 0 \end{bmatrix}$$

(e) Population studies

Ordinary differential equations arise in the analysis of simple models of the growth of populations. We denote by $p(t)$ the number of inhabitants of a prescribed area at time t. The time rate of change of $p(t)$ depends on:

(i) the number of individuals born in that time; we assume that the birth rate is Np, where N may depend on both p and t:
(ii) the number of individuals dying; we assume that the death rate is Mp, where M may be a function of p and t:
(iii) the rate at which individuals enter the area; this is taken as $I(p, t)$:
(iv) the rate $E(p, t)$ at which individuals leave the area.

Making the balance we obtain the first-order differential equation

$$\dot{p} = (N - M)p + (I - E) \tag{1.22}$$

In one population model it is assumed that I and E are constants and that

$$N = n - \nu p, \qquad M = m + \mu p, \tag{1.23}$$

where m, n, μ and ν are constants. Substituting from equations (1.23) into equation (1.22) we obtain the differential equation

$$\dot{p} = \varepsilon p - kp^2 + (I - E). \tag{1.24}$$

This first-order differential equation is known as *Verhulst's equation*; the constant ε is called the *coefficient of increase* and k is called the *limiting coefficient*.

If the population is isolated (i.e. if there is neither immigration nor emigration), $I = E = 0$ and equation (1.24) reduces to the simpler form

$$\dot{p} = kp(\omega - p), \qquad \omega = \varepsilon/k \tag{1.25}$$

(f) Differential equations in epidemiology

Ordinary differential equations arise in a similar way in the theory of epidemics.

In a model for a simple epidemic it is assumed that infection spreads between members of the community, but that there is no removal from circulation. Suppose that we consider a community of susceptibles into which a single infective is introduced. We shall denote by x and y the number of susceptibles and infectives, respectively, at any time t, so that $x+y=n+1$, the total population size. We suppose that the whole population is subject to some process of homogeneous mixing, so that the number of new infections occurring in time $\mathrm{d}t$ is $\beta xy\,\mathrm{d}t$, where β is the infection rate. Then

$$\dot{x}=-\beta xy.$$

Substituting $y=n+1-x$ into this equation, we find that the time variation of x is given by the differential equation

$$\dot{x}=-\beta x(n+1-x), \tag{1.26}$$

with initial condition $x=n$ at time $t=0$.

We now consider a population of a total of n individuals comprising, at time t, x susceptibles, y infectives in circulation and z individuals who are isolated, dead or recovered and immune. Thus $x+y+z=n$. If β and γ are the infection rate and removal rate respectively, then we have the three simultaneous differential equations

$$\dot{x}=-\beta xy, \qquad \dot{y}=\beta xy-\gamma y, \qquad \dot{z}=\gamma y. \tag{1.27}$$

If, furthermore, we introduce a birthrate parameter μ, so as to give $\mu\,\mathrm{d}t$ new susceptibles in time $\mathrm{d}t$, we then have the modified equations

$$\dot{x}=-\beta xy+\mu, \qquad \dot{y}=\beta xy-\gamma y. \tag{1.28}$$

(g) Electric circuits

The simplest differential equation in circuit theory arises when a current flows through a resistance R and a self-inductance L (cf Fig. 3a). If the applied voltage is denoted by $E(t)$, the differential equation for the current x is

$$L\dot{x}+Rx=E(t), \tag{1.29}$$

so that the current in such a simple circuit is determined by an ordinary differential equation of the first order. On the other hand, if the circuit contains a capacitance C (see Fig. 3b) and if $Q(t)$ is the charge on the plates at time t, we have the equation

$$L\dot{x}+Rx+\frac{Q}{C}=E(t);$$

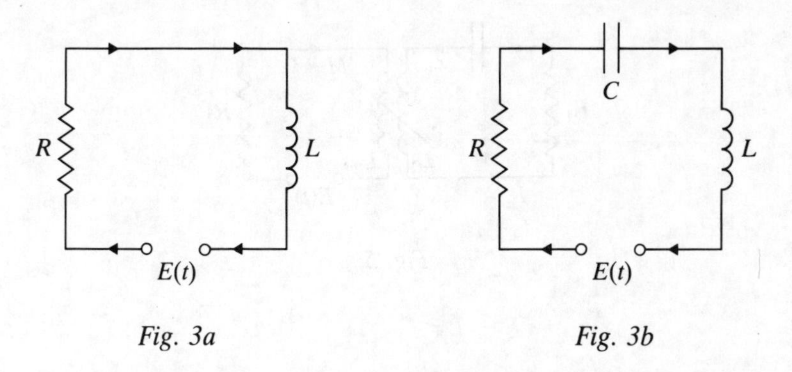

Fig. 3a *Fig. 3b*

the relation between the current $x(t)$ and the charge on the plates is given by $x(t)=\dot{Q}(t)$, so that the current in the circuit is the time derivative of the solution of the second-order ordinary differential equation

$$L\ddot{Q}+R\dot{Q}+Q/C=E(t). \tag{1.30}$$

The situation is much more complicated when we have two coupled circuits (see Fig. 4).

If two circuits with resistances R_1, R_2, capacitances C_1, C_2 and self-inductances L_1, L_2 respectively are coupled through a mutual inductance M, then the equations governing the currents x_1 and x_2 in the two circuits are:

$$L_1\dot{x}+M\dot{x}_2+R_1x_1=E_1(t)$$

$$L_2\dot{x}_2+M\dot{x}_1+R_2x_2=E_2(t).$$

In these equations the functions E_1 and E_2 are time-variable functions of the applied voltage. In this case the electrical network is described by a pair of first-order linear equations.

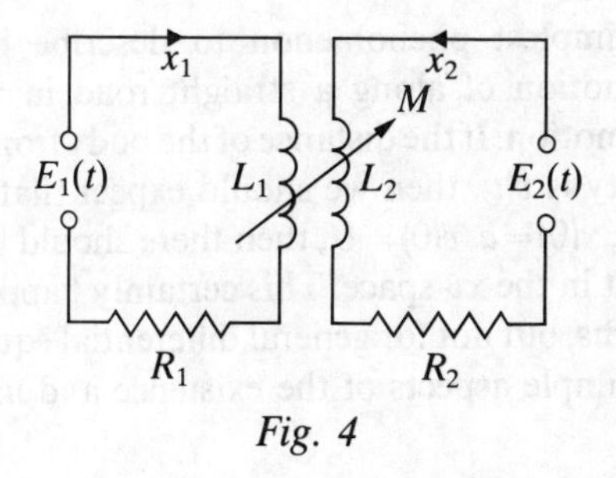

Fig. 4

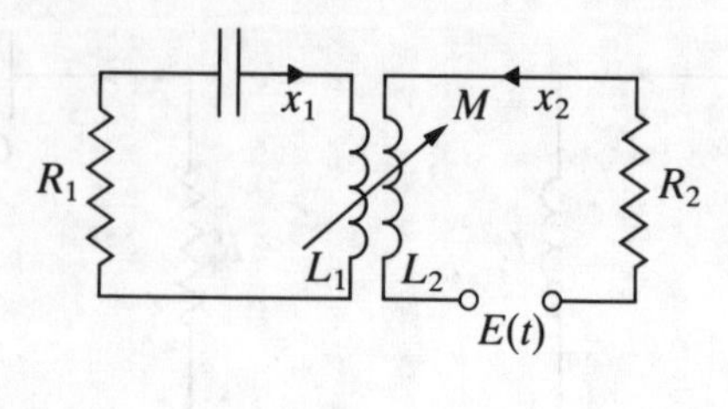

Fig. 5

On the other hand if we consider the pair of coupled circuits shown in Fig. 5 we find that the distribution of current in the network is described by the pair of equations

$$M\dot{x}_1 + L_2\dot{x}_2 + R_2 x_2 = E(t) \tag{1.31}$$

$$L_1\dot{x}_1 + R_1\dot{x}_1 + \frac{Q}{C} + M\dot{x}_2 = 0;$$

combining these equations with the relation between current and charge $\dot{Q} = x_1$, we obtain the pair of equations (1.31) and

$$L_1\ddot{x}_1 + R_1\dot{x}_1 + x_1/C + M\ddot{x}_2 = 0.$$

Other combinations of electrical elements can, of course, lead to different circuits and therefore to different systems of differential equations.

In this first section we have illustrated how certain physical and geometrical phenomena may have their essential features expressed in terms of ordinary differential equations. The process of elimination by which a differential equation may be obtained from a primitive is called elimination; the action of recovering the primitive or an equivalent expression is known as the *integration* – or merely the *solution* – of the differential equation.

Perhaps the simplest phenomenon to describe by a differential equation is the motion of along a straight road in accordance with Newton's laws of motion. If the distance of the body from a fixed origin is $x(t)$ and its velocity is $v(t)$, then we should expect that if, at an instant measured by $t = 0$, $x(0) = a$, $v(0) = U$, then there should be a unique path through that point in the xv-space. This certainly happens in almost all dynamical problems, but not for general differential equations. We shall consider certain simple aspects of the existence and uniqueness in the next section.

§2 Uniqueness and existence

The basic question we ask is the following: Does there exist a solution for the initial-value problem posed by the equations

$$y'(x) = f(x, y), \tag{2.1}$$

$$y(a) = b, \tag{2.2}$$

where $f(x, y)$ is a prescribed function of the two real variables in some region of the xy-plane containing the point (a, b)? In other words, which regions – if any – of the xy-plane have the property that through each point of the region there passes a unique curve corresponding to the solution of the first-order equation (2.1).

For instance, it is easily shown that the solution of the boundary-value problem

$$y'(x) = \begin{cases} (1-2x)y, & x \geqslant 0 \\ (2x-1)y, & x \leqslant 0 \end{cases}$$

$$y(0) = 0,$$

can be written in the form

$$y(x) = \begin{cases} e^{x-x^2}, & x \geqslant 0, \\ e^{x^2-x}, & x \leqslant 0. \end{cases}$$

It is obvious from these expressions – as from the accompanying graph Fig. 6 – that there is only one solution curve passing through the point (0, 1). In other words, the initial boundary problem has a *unique* solution.

On the other hand there are *two* solutions of the equation

$$y' = y^{1/2}$$

passing through the origin. The first solution is the x-axis, but there is another one. The second one is easily verified to be

$$y = \begin{cases} \tfrac{1}{4}x^2 & x \geqslant 0, \\ -\tfrac{1}{4}x^2 & x \leqslant 0 \end{cases}$$

as shown in Fig. 7.

A further example is shown by the more complicated form for the function

$$f(x, y) = \begin{cases} \dfrac{4x^3 y}{x^4 + y^2}, & x \neq 0,\ y \neq 0, \\ 0, & x = y = 0 \end{cases}$$

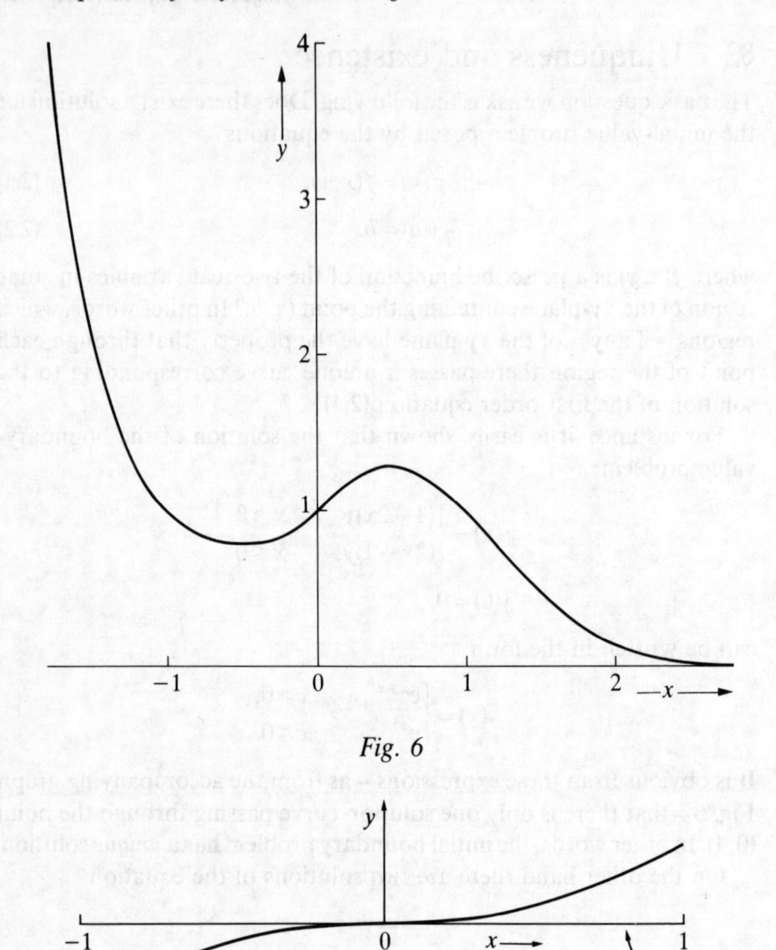

Fig. 6

Fig. 7

occurring in equation (2.1). It is easily verified that the function

$$y(x)=c^2-\sqrt{(x^4+c^4)}$$

satisfies this equation – and passes through the origin – for each real value of the parameter c. In this case, therefore, there is an infinity of curves satisfying the differential equation and the boundary condition imposed at the origin. (Cf. Fig. 8.)

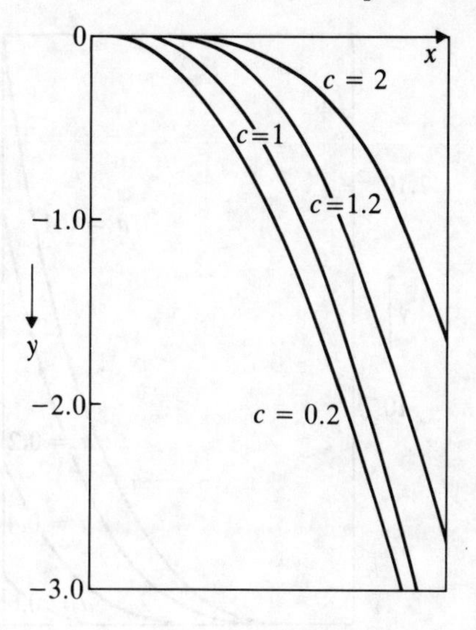

Fig. 8

Finally, we consider the ordinary differential equation

$$y' = y^{2/3}, \tag{2.3}$$

satisfying $y(0) = 0$. It will be noted that $y = \frac{1}{27}(x + c)^3$ is a solution, as is the null function $y(x) = 0$, for all real values of x. Suppose now that a is a real number $0 < a < 1$; then the function

$$y(x) = \begin{cases} 0, & 0 \leqslant x \leqslant a \\ \frac{1}{27}(x - a)^3, & a \leqslant x \leqslant 1 \end{cases}$$

belongs to $C^2[0, 1]$ and satisfies (2.3) and $y(0) = 0$. There are infinitely many such solutions each of the shape shown in Fig. 9.

Similarly, if $c > 3$ the functions

$$y(x) = \begin{cases} \frac{1}{27}(x - 3)^3, & 0 \leqslant x \leqslant 3 \\ 0, & 3 \leqslant x \leqslant a \\ \frac{1}{27}(x - a)^3, & a < x \leqslant c \end{cases}$$

are solutions of (2.3), $y(0) = -1$. Hence if $c > 3$ there are infinitely many solutions of this boundary-value problem, each of them of the shape shown in Fig. 10. However if $c \leqslant 3$ this initial-value problem has the *unique* solution

$$y(x) = \tfrac{1}{27}(x - 3)^2, \quad 0 \leqslant x \leqslant c.$$

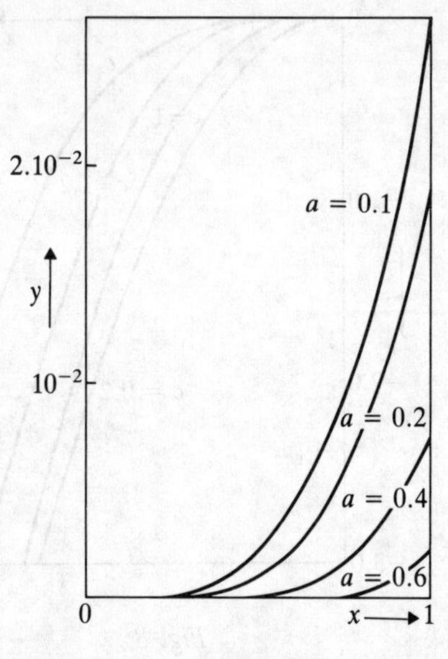

Fig. 9

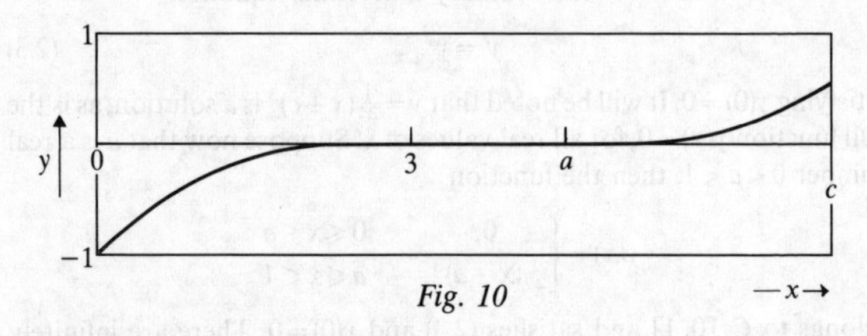

Fig. 10

These examples of the simple equation (2.1) show that even when a solution of a boundary-value problem exists, it may not be unique. The function f has to be continuous at (a, b) if the solution of the boundary-value problem is to exist, but something more is needed for uniqueness.

Definition. The function $f(x, y)$ is said to satisfy a *Lipschitz condition* with respect to y in a region D of the xy-plane, if there exists a constant $K \geqslant 0$ such that

$$|f(x, y_1) - f(x, y_2)| \leqslant K|y_1 - y_2|, \qquad y_1, y_2 \in D$$

The constant K is called a *Lipschitz constant*, and we write $f \in \text{Lip}_y(D)$.

The problem of solving the differential equation (2.1) subject to the initial condition (2.2) is obviously equivalent to solving the integral equation

$$y = b + \int_a^x f(t, y(t))\,\mathrm{d}t. \tag{2.4}$$

The fundamental existence theorem is proved by establishing a sequence of functions $\{y_n\}$ by the formula

$$y_{n+1}(x) = b + \int_a^x f(t, y_n(t))\,\mathrm{d}t \tag{2.5}$$

and then showing that the limit of the sequence as $n \to \infty$ tends to the unique solution of the boundary-value problem.

Theorem 2.1 *If $f \in C(D)$, $f \in \mathrm{Lip}_y(D)$, where D is an open connected set in the xy-plane and if $(a, b) \in D$, the initial-value problem*

$$y'(x) = f(x, y), \qquad y(a) = b,$$

has a unique solution on some real interval containing $x = a$.

That the continuity of $f(x, y)$ is sufficient to guarantee existence but not uniqueness is illustrated by the further boundary-value problem:

Example 2.1

$$y' = y^{1/3}, \qquad y(0) = 0.$$

It is easily verified that the solution of this equation is $y = \pm(\frac{2}{3}x)^{3/2}$ so that there are *two* branches through the origin. Uniqueness breaks down because $f(x, y) = y^{1/3}$ does not satisfy a Lipschitz condition in the neighbourhood of the origin. To establish this we need only produce a suitable pair of points for which a relation of the type

$$|f(x, y_1) - f(x, y_2)| \leqslant K|y_1 - y_2|$$

fails to hold for a suitable K. If we take the points $(x, 0)$, (x, y_1), then

$$|f(x, y_1) - f(x, 0)| = |y_1|^{1/3}$$

and so

$$\frac{|f(x, y_1) - f(x, 0)|}{|y_1|} = |y_1|^{-2/3}.$$

By choosing y_1 sufficiently small it is clear that we can make $|y_1|^{-2/3}$ larger than any preassigned constant. Therefore, f cannot belong to

Lip (Ω), where Ω is any neighbourhood of the origin – although $f \in C(\Omega)$.

Since the number b occurring in Theorem **2.1** is arbitrary we see that the general solution of a first-order differential equation involves a single arbitrary constant.

Now if we consider the initial-value problem for an ordinary differential equation of the second order, say

$$y'' + a(x)y' + x^2y^2 = 0,$$

we could write this in component form

$$y_1(x) = y(x)$$
$$y_2(x) = y'(x)$$

as

$$y_1' = y_2$$
$$y_2' = -ay_2 - x^2y_1^2.$$

Writing $\mathbf{y}$ for the two-vector

$$\begin{bmatrix} y_1 \\ y_2 \end{bmatrix}$$

we can formulate the initial value problem in the vector form

$$\mathbf{y}' = \mathbf{f}(x, \mathbf{y}), \qquad \mathbf{y}(0) = \mathbf{y}^0$$

In a similar way the system of equations

$$y_1' = f_1(x, y_1, \ldots, y_n)$$
$$y_2' = f_2(x, y_1, \ldots, y_n)$$
$$\vdots \qquad \vdots$$
$$y_n' = f_n(x, y_1, \ldots, y_n)$$

$$y_1(0) = b_1, \quad y_2(0) = b_2, \quad \ldots, \quad y_n(0) = b_n$$

may be written in the form

$$\mathbf{y}' = \mathbf{f}(x, \mathbf{y}) \qquad \mathbf{y}(0) = \mathbf{b}$$

for the vectors $\{y_1, y_2, \ldots, y_n\}$ and $\{f_1, f_2, \ldots, f_n\}$.

We have to reformulate the Lipschitz condition in this case. If D is an open connected set $\mathbb{R}^{n+1}$ we say that $\mathbf{f} \in \text{Lip}_{\mathbf{y}}\,(D)$ if there exists a positive constant K such that

$$\|\mathbf{f}(x, \mathbf{y}_1) - \mathbf{f}(x, \mathbf{y}_2)\| \leqslant K\|\mathbf{y}_1 - \mathbf{y}_2\|, \qquad \mathbf{y}_1, \mathbf{y}_2 \in D.$$

If $\mathbf{y}$ is a vector in $\mathbb{R}^n$, then we may take

$$\|\mathbf{y}\| = \sum_{j=1}^{n} \{|y_j|^2\}^{1/2}$$

as the norm of $\mathbf{y}$; an equally suitable norm is

$$\|\mathbf{y}\| = \sum_{j=1}^{n} |y_j|,$$

or alternatively

$$\|\mathbf{y}\| = \max_j |y_j|.$$

The basic theorem is:

Theorem 2.2 *If $f \in C(\Omega)$, $f \in \mathrm{Lip}_y(\Omega)$, where Ω is an open connected set in $\mathbb{R} \times \mathbb{R}^n$ and if $(x_0, \mathbf{b}) \in \Omega$, the initial-value problem*

$$\mathbf{y}' = \mathbf{f}(x, \mathbf{y}), \qquad \mathbf{y}(x_0) = \mathbf{b} \tag{2.6}$$

has a unique solution on some real interval containing x_0.

Since the initial value $\mathbf{b}$ is an arbitrary element in $\mathbb{R}^n$, we deduce that the general solution involves n arbitrary real constants.

Suppose now that the vector-valued function $\mathbf{f}$ in equation (2.6) satisfies the hypotheses of Theorem 2.2, and that $\mathbf{z}(x)$ is a solution whose graph

$$C = \{(x, \mathbf{z}(x)); a < x < b\}$$

lies entirely in the open connected region Ω of the $(x, \mathbf{y})$-space. We now define

$$U_{z\delta} = \{(x, \mathbf{z}(x)): a < x < b, \|\mathbf{y} - \mathbf{z}\| < \delta\}$$

Fig. 11 shows $U_{z\delta}$ in a typical case in E_2. Obviously

$$\mathbf{z} \in T = \{\mathbf{y}: y_i \in C[a, b], i = 1, 2, \ldots, n\}.$$

A norm for this space is provided by the definition

$$\|\mathbf{y}\| = \sup_{t \in [a,b]} \|\mathbf{y}(t)\|$$

and the spherical neighbourhood

$$K(\mathbf{z}, \delta) = \{\mathbf{y}: \mathbf{y} \in T, \|\mathbf{y} - \mathbf{z}\| < \delta\}$$

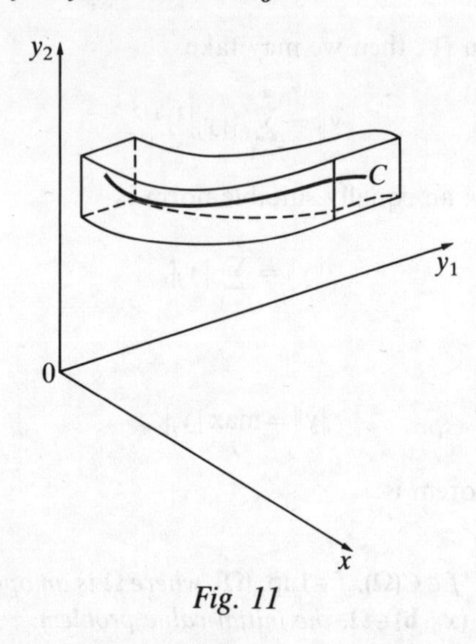

Fig. 11

contains precisely those elements of T whose graphs are in the region $U_{z\delta}$. It should be noticed that if δ is sufficiently small the region $U_{z\delta}$ lies entirely in Ω. Hence we may choose a positive real number δ_1 such that $U_{z\delta} \subset \Omega$ if $\delta < \delta_1$. Intuitively we should suppose that any spherical neighbourhood of $\mathbf{z}$ would contain solutions of (2.6) other than $\mathbf{z}$ itself. That this is so is proved in:

Theorem 2.3 *If* $\mathbf{z}$ *is a solution of* $\dot{\mathbf{y}} = \mathbf{f}(x, \mathbf{y})$ *then, corresponding to any* $\varepsilon > 0$, *there exists a* $\delta > 0$ *such that the solution for which* $(\boldsymbol{\xi}, \mathbf{b})$ *is an initial point is in the spherical neighbourhood* $K(\mathbf{z}, \varepsilon)$ *if* $(\boldsymbol{\xi}, \mathbf{b}) \in U_{z\delta}$.

The solution of (2.6) may be written $\mathbf{z}(x, x_0, \mathbf{b})$. A point $(x_0, \mathbf{b}) \in \Omega$ which is used to determine the solution is called an *initial point* of that solution. The question that now arises is whether or not a solution depends continuously on its initial point? It turns out that this question has an affirmative answer. Also since the solution $\mathbf{z}(x, x_0, \mathbf{b})$ may be regarded as a function of the $n+2$ variables $x, x_0, b_1, \ldots, b_n$ we can say that the solution $\mathbf{z}(x, x_0, b_1, \ldots, b_n)$ is a continuous function of its $n+2$ variables.

The remainder of this chapter will be devoted to equations of the first order and first degree; that is, to equations that may be written in the alternative forms

$$P(x, y) + Q(x, y)y' = 0, \qquad (2.7)$$

$$P(x, y)\,dx + Q(x, y)\,dy = 0, \tag{2.8}$$

where P and Q do **not** involve the derivative y'.

Since from any primitive equation of the form

$$u(x, y) = c, \tag{2.9}$$

where c is an arbitrary constant, we deduce that

$$\frac{\partial u}{\partial x}\,dx + \frac{\partial u}{\partial y}\,dy = 0,$$

it follows that a necessary condition for (2.9) to be an integral of (2.7) is

$$P\frac{\partial u}{\partial y} - Q\frac{\partial u}{\partial x} = 0. \tag{2.10}$$

§3 Separation of variables

Among equations of the type $P + Qy' = 0$ the simplest are those in which P is a function of x alone and Q of y alone, say $P = M(x)$, $Q = N(y)$. The general integral is obtained by direct integration, thus:

$$\int M(x)\,dx + \int N(y)\,dy = C,$$

where C is an arbitrary constant of integration.

More generally, let P and Q be the products of a term involving x but not y and a term involving y but not x, so that the equation has the form

$$M(x)R(y) + N(y)S(x)y' = 0.$$

The variables are then said to be *separable*, for on division by $R(y)S(x)$ the equation becomes

$$\frac{M(x)}{S(x)} + \frac{N(y)}{R(y)}\,y' = 0$$

and has the general integral

$$\int \frac{M(x)}{S(x)}\,dx + \int \frac{N(y)}{R(y)}\,dy = C.$$

When, as in the above cases, the process leads to an expression that involves integral signs the result is said to be an integration by quadratures. This implies that the problem has been reduced from one in differential equations to an equivalent one in the integral calculus. If it is found impossible to evaluate one or other of the integrals, an explicit solution of the equation is impossible, and the solution by quadratures must be regarded as the best attainable, unless an alternative line of approach can be discovered.

Example 3.1

$$x(y^2-1)-y(x^2-1)y'=0.$$

Separating the variables:

$$\frac{x}{x^2-1}-\frac{yy'}{y^2-1}=0.$$

Integrating,

$$\log|x^2-1|-\log|y^2-1|=C$$

or

$$\log\left|\frac{x^2-1}{y^2-1}\right|=-\log c$$

which may be written

$$(y^2-1)=c(x^2-1).$$

(Notice that replacing the arbitrary constant C by another arbitrary form, as in this case $C=-\log c$, may help to simplify the general integral.)

Example 3.2

$$(1-x^2)^{-1/2}+(1-y^2)^{-1/2}y'=0.$$

The variables are separate; direct integration gives

$$\text{arc}\,\{\sin x\}+\text{arc}\,\{\sin y\}=C$$

which is the general integral. This may be transformed into an equivalent expression by slightly rearranging the terms and taking the sine of both members, thus:

$$\sin\{\text{arc}\sin y\}=\sin\{C-\text{arc}\,(\sin x)\}$$

or, since

$$\cos\{\text{arc}\sin x\}=\pm\sqrt{(1-x^2)},$$

$$y=\pm\sqrt{(1-x^2)}\sin C-x\cos C$$

and, rationalizing,

$$(y+x\cos C)^2=(1-x^2)\sin^2 C,$$

i.e.

$$x^2+y^2+2xy\cos C=\sin^2 C$$

or, if $c=\cos C$,

$$x^2+y^2+2cxy=1-c^2.$$

Example 3.3 A change in variables may sometimes succeed in converting an equation into another with separate variables. For instance, in

$$(x+y)+y'=0$$

the variables are not separable, but if y is replaced by $v-x$, the equation is transformed into

$$(v-1)+v'=0.$$

The variables x and v are separable, thus

$$1+\frac{v'}{v-1}=0$$

which leads to

$$x+\log|v-1|=\log c \qquad \text{or} \qquad (v-1)\,e^x=c$$

so that the original equation has a general integral of the form

$$(x+y-1)\,e^x=c \qquad \text{or} \qquad y=c\,e^{-x}-x+1.$$

The most important instance of reduction by change of variables occurs in the case of the homogeneous equation which now follows.

§4 The homogeneous type

The equation

$$P(x,y)+Q(x,y)y'=0$$

is said to be of homogeneous type when P and Q are homogeneous functions of x and y of the same degree. If the degree is m, the substitution $y=vx$ will reduce P and Q to the forms

$$P(x,vx)=x^mR(v), \qquad Q(x,vx)=x^mS(v),$$

where R and S are independent of x. Thus the factor x^m may be cancelled out of the equation, which becomes

$$R(v)+S(v)\{v+xv'\}=0$$

or

$$\{R(v)+vS(v)\}+xS(v)v'=0.$$

Separating the variables and integrating, we have

$$\int\frac{S(v)\,\mathrm{d}v}{R(v)+vS(v)}+\log x=C,$$

and when the integral in v has been evaluated, the substitution $v=y/x$ will give the general integral of the original equation.

Example 4.1

$$(x^2-y^2)+2xyy'=0.$$

The two terms in this equation are homogeneous and of the second degree in x and y; the above process is therefore applicable. Making the substitution mentioned, we have

$$x^2(1-v^2)+2x^2v(v+xv')=0;$$

x^2 cancels out leaving

$$1+v^2+2xvv'=0.$$

Separating the variables and integrating we have the solution

$$\int\frac{2v\,\mathrm{d}v}{1+v^2}+\int\frac{\mathrm{d}x}{x}=C,$$

or

$$\log(1+v^2)+\log x=\log c,$$

i.e.

$$(1+v^2)x=c,$$

which leads to the general integral

$$x^2+y^2=cx.$$

Example 4.2

$$(2y\,\mathrm{e}^{y/x}-x)y'+2x+y=0.$$

Here each term is of the first degree in x and y, for $\mathrm{e}^{y/x}$ is of degree zero.

Writing $y = vx$, $y' = xv' + v$ and cancelling x, we obtain the equation

$$(2v\,\mathrm{e}^v - 1)xv' + 2(v^2\,\mathrm{e}^v + 1) = 0.$$

Separating the variables, we see that this reduces to

$$\frac{(2v\,\mathrm{e}^v - 1)v'}{v^2\,\mathrm{e}^v + 1} + \frac{2}{x} = 0.$$

Integrating, we have

$$\log(v^2 + \mathrm{e}^{-v}) + 2\log x = C,$$

whence the general integral

$$y^2 + x^2\,\mathrm{e}^{-y/x} = c.$$

§5 The equation with linear coefficients

Although the equation

$$(ax + by + c) + (a'x + b'c + c')y' = 0 \tag{5.1}$$

is not of homogeneous type, it may be reduced to that type by a simple substitution. The equations

$$ax + by + c = 0, \qquad a'x + b'y + c' = 0 \tag{5.2}$$

represent a pair of straight lines which will intersect unless the condition for parallelism, i.e. $a'/a = b'/b$ or $ab' - a'b = 0$, is satisfied. Suppose that (h, k) is the point of intersection; transferring the origin to that point by the substitution

$$x = h + X, \qquad y = k + Y$$

we find that the equation becomes

$$(aX + bY) + (a'X + b'Y)Y' = 0.$$

It is now homogeneous; the substitution $Y = vX$ followed by separation of variables leads to the general integral

$$\log CX + \int \frac{(a' + b'v)\,\mathrm{d}v}{a + (a' + b)v + v'v^2} = 0,$$

whose ultimate form depends upon whether the roots of the denominator of the integrand are real distinct, coincident or complex, i.e. according as $(a' + b)^2$ is greater than, equal to, or less than $4ab'$.

In the exceptional case when the lines (5.2) are parallel, that is when

$a'/a = b'/b = k$ (say), the equation can be written

$$(ax+by+c)+\{k(ax+by)+c'\}y'=0$$

Taking $z=ax+by$ as a new variable to replace y, we obtain the equation

$$b(z+c)+(kz+c')(z'-a)=0.$$

Separating the variables and integrating, we have

$$x+\int\frac{(kz+c')\,\mathrm{d}z}{(b-ak)z+bc-ac'}=const.$$

The above equations are particular cases of an equation of the type

$$y'=F\left(\frac{ax+by+c}{a'x+b'y+c'}\right)$$

which may be reduced to a form integrable by quadratures by the same process.

Example 5.1

$$y'=\frac{4x-y+7}{2x+y-1}.$$

The lines $4x-y+7=0$, $2x+y-1=0$ meet in $(-1, 3)$; writing $x=X-1$, $y=Y+3$ we have

$$(2X+Y)Y'=(4X-Y).$$

The transformation $Y=vX$ reduces this equation to

$$(2+v)Xv'+(v^2+3v-4)=0$$

which becomes, on separation of the variables and formation of partial fractions,

$$\left\{\frac{3}{v-1}+\frac{2}{v+4}\right\}\frac{\mathrm{d}v}{\mathrm{d}X}+\frac{5}{X}=0.$$

Integrating this equation we obtain the solution

$$3\log|v-1|+2\log|v+4|+5\log|X|=C,$$

i.e.

$$(v-1)^3(v+4)^2X^5=c$$

or

$$(Y-X)^3(Y+4X)^2=c.$$

Reverting to the variables x, y we have the general integral

$$(y-x-4)^3(y+4x+1)^2=c.$$

Example 5.2

$$(2x-4y+5)y'+x-2y+3=0.$$

If $z=x-2y$, $2y'=1-z'$; with this transformation the equation becomes

$$(2z+5)z'=4z+11.$$

Separating the variables, we obtain

$$\left(1-\frac{1}{4z+11}\right)z'=2$$

whence we have

$$4z-\log|4z+11|=8x-C,$$

giving the general integral

$$4x+8y+\log|4x-8y+11|=C.$$

§6 Exact equations

When the primitive of a differential equation involves the arbitrary constant C explicitly, as:

$$u(x, y)=C, \tag{6.1}$$

the operation of taking the differential eliminates C automatically, thus:

$$\frac{\mathrm{d}u(x, y)}{\mathrm{d}x}=0 \tag{6.2}$$

or

$$\frac{\partial u}{\partial x}+\frac{\partial u}{\partial y}y'=0. \tag{6.3}$$

Conversely, if a differential equation of the form

$$P(x, y)+Q(x, y)y'=0 \tag{6.4}$$

has originated in such a process, and if no variable factor has been cancelled out, it must be equivalent to one of the form (6.3) and thus to

(6.2), and therefore it must possess a general integral of the form (6.1). Such an equation is said to be *exact*.

Thus, in order that (6.4) may be exact, there must exist a function $u(x, y)$ such that

$$P(x, y)=\frac{\partial u}{\partial x}, \qquad Q(x, y)=\frac{\partial u}{\partial y}. \tag{6.5}$$

and therefore

$$\frac{\partial P}{\partial y}=\frac{\partial^2 u}{\partial x\,\partial y}=\frac{\partial Q}{\partial x}.$$

Hence the theorem: *for the differential equation* (6.4) *to be exact it is necessary that* $P(x, y)$ *and* $Q(x, y)$ *be linked by the relation*

$$\frac{\partial P}{\partial y}=\frac{\partial Q}{\partial x} \tag{6.6}$$

This is known as the condition of integrability,† for when it is satisfied the primitive $u(x, y)=C$ can be recovered by the following process. Starting from the relation

$$\frac{\partial u(x, y)}{\partial x}=P(x, y),$$

integrating and remembering that since the differentiation with respect to x was partial, the inverse process of integration will introduce, as the arbitrary element, a function of y, we have

$$\begin{aligned} u(x, y) &= \int P(x, y)\,\mathrm{d}x+f(y) \\ &= S(x, y)+f(y) \qquad \text{(say).} \end{aligned}$$

But, on the other hand,

$$Q(x, y)=\frac{\partial u}{\partial y}=\frac{\partial S}{\partial y}+f'(y),$$

an equation which will give $f'(y)$ since Q and S are both known; the final integration to obtain $f(y)$ will introduce the arbitrary constant C of the general integral.

† Immediate integrability is implied; when the condition is not satisfied, the equation is still integrable, though not without some preliminary manipulation.

Example 6.1

$$\frac{(1+y^2)y+(1+x^2)xy'}{(1+x^2+y^2)^{3/2}}=0.$$

Since, in the above notation,

$$\frac{\partial P}{\partial y}=\frac{\partial}{\partial y}\left\{\frac{y+y^3}{(1+x^2+y^2)^{3/2}}\right\}=\frac{1+x^2+y^2+3x^2y^2}{(1+x^2+y^2)^{5/2}}=$$

$$=\frac{\partial}{\partial x}\left\{\frac{x+x^3}{(1+x^2+y^2)^{3/2}}\right\}=\frac{\partial Q}{\partial x}$$

the equation is exact. Hence we are entitled to write

$$u(x, y)=\int\frac{(1+y^2)y\,\mathrm{d}x}{(1+x^2+y^2)^{3/2}}+f(y)$$

$$=\int\frac{y\,\mathrm{d}x}{(1+x^2+y^2)^{1/2}}-\int\frac{x^2y\,\mathrm{d}x}{(1+x^2+y^2)^{3/2}}+f(y)$$

$$=\int\frac{y\,\mathrm{d}x}{(1+x^2+y^2)^{1/2}}+\int xy\frac{\partial}{\partial x}\left\{\frac{1}{(1+x^2+y^2)^{1/2}}\right\}\mathrm{d}x+f(y)$$

$$=\frac{xy}{\sqrt{(1+x^2+y^2)}}+f(y),$$

on integrating the second integral by parts. The equation

$$\frac{\partial u}{\partial y}=\frac{x+x^3}{(1+x^2+y^2)^{3/2}}+f'(y)$$

shows that $f'(y)$ is zero, or $f(y)$ is a constant. The general integral is therefore

$$\frac{xy}{\sqrt{(1+x^2+y^2)}}=C.$$

Note that although the given equation is exact as it stands, it would cease to be exact if the denominator of the left-hand member were removed.

Example 6.2

$$\log(y^2+1)+\frac{2y(x-1)}{y^2+1}\,y'=0.$$

The condition for integrability is satisfied, therefore we write

$$u(x, y)=\int \log (y^2+1)\,\mathrm{d}x+f(y)$$
$$=x \log (y^2+1)+f(y).$$

From this we obtain

$$\frac{\partial u}{\partial y}=\frac{2xy}{y^2+1}+f'(y);$$

comparison with the coefficient of y' in the given equation shows that

$$f'(y)=-\frac{2y}{y^2+1} \qquad \text{or} \qquad f(y)=-\log (y^2+1)+const.$$

The general integral therefore is

$$(x-1)\log (y^2+1)=C.$$

§7 Integrating factors

It will now be supposed that the condition for integrability (6.6) is not satisfied. This being so, an explicit general integral $u(x, y)=C$ is not immediately obtainable, but, on the other hand, the theory of the existence of integrals is able to establish the fact that a general integral in which the constant C is implicit does exist. This may be interpreted to mean that, given any non-exact equation

$$P+Qy'=0, \tag{7.1}$$

there always exists an integrating factor $\mu(x, y)$ such that the modified equation

$$\mu P+\mu Q y'=0 \tag{7.2}$$

is eact. Assuming that this is so, $\mu(x, y)$ must satisfy identically the relation

$$\frac{\partial(\mu P)}{\partial y}=\frac{\partial(\mu Q)}{\partial x} \tag{7.3}$$

or

$$P\frac{\partial \mu}{\partial y}-Q\frac{\partial \mu}{\partial x}+\mu\left\{\frac{\partial P}{\partial y}-\frac{\partial Q}{\partial x}\right\}=0. \tag{7.4}$$

To find a general solution of this partial differential equation is a much more difficult problem than that of solving the ordinary differential equation originally proposed. For present purposes, however, any particular solution of (7.4) would suffice, and such a solution may frequently be obtained by trial. When any integrating factor μ, satisfying (7.4), has been obtained and introduced into (7.2), this equation becomes exact and may be integrated by the method of §6. The limitations of this method of integration are obvious; nevertheless there are particular cases which may be treated systematically; these will be indicated in §9.

Example 7.1 The equation

$$y - xy' = 0$$

is not exact. Any integrating factor μ will satisfy

$$\frac{\partial(y\mu)}{\partial y} = -\frac{\partial(x\mu)}{\partial x}$$

or

$$x\frac{\partial\mu}{\partial x} + y\frac{\partial\mu}{\partial y} + 2\mu = 0.$$

It can be verified that

$$\frac{1}{x^2}, \quad \frac{1}{y^2}, \quad \frac{1}{xy}$$

are among the possible values of μ. Taking $\mu = 1/x^2$ we have

$$\frac{y}{x^2} - \frac{1}{x}y' = 0$$

which is exact and has the solution $y/x = C$. Similarly $\mu = 1/y^2$ leads to the equivalent solution $x/y = C'$. Taking $\mu = 1/xy$ we separate the variables:

$$\int\frac{dx}{x} - \int\frac{dy}{y} = 0$$

or

$$\log x - \log y = const.,$$

whence $x/y = const.$

Example 7.2 $(1+x^2+y^2)^{-3/2}$ is an integrating factor for

$$(1+y^2)y+(1+x^2)xy'=0$$

(cf. Example **6.1**). Another integrating factor is the reciprocal of $xy(1+x^2)(1+y^2)$; this separates the variables.

§8 The quotient of two integrating factors

Suppose that μ_1 and μ_2 are two integrating factors of

$$P+Qy'=0, \tag{8.1}$$

whose ratio is not of itself a constant. Then the equation

$$\mu_2/\mu_1=C \qquad \text{or} \qquad \mu_2=C\mu_1 \tag{8.2}$$

is an integral of (8.1). For (8.2) is a primitive of the differential equation

$$\left\{\mu_2\frac{\partial\mu_1}{\partial x}-\mu_1\frac{\partial\mu_2}{\partial x}\right\}+\left\{\mu_2\frac{\partial\mu_1}{\partial y}-\mu_1\frac{\partial\mu_2}{\partial y}\right\}y'=0. \tag{8.3}$$

But since μ_1 and μ_2 are integrating factors of (8.1)

$$P\frac{\partial\mu_1}{\partial x}-Q\frac{\partial\mu_1}{\partial x}+\mu_1\left\{\frac{\partial P}{\partial y}-\frac{\partial Q}{\partial x}\right\}=0, \tag{8.4}$$

$$P\frac{\partial\mu_2}{\partial y}-Q\frac{\partial\mu_2}{\partial x}+\mu_2\left\{\frac{\partial P}{\partial y}-\frac{\partial Q}{\partial x}\right\}=0, \tag{8.5}$$

and hence

$$\mu_2\left\{P\frac{\partial\mu_1}{\partial y}-Q\frac{\partial\mu_1}{\partial x}\right\}-\mu_1\left\{P\frac{\partial\mu_2}{\partial y}-Q\frac{\partial\mu_2}{\partial x}\right\}=0,$$

i.e.

$$\left\{\mu_2\frac{\partial\mu_1}{\partial y}-\mu_1\frac{\partial\mu_2}{\partial y}\right\}P=\left\{\mu_2\frac{\partial\mu_1}{\partial x}-\mu_1\frac{\partial\mu_2}{\partial x}\right\}Q,$$

which reduces (8.3) to (8.1), thereby proving the theorem.

This theorem is equivalent to the statement that if one integrating factor of (8.1) is known, an unlimited number of others may be obtained. Thus if μ_1 is known and we write $\mu_2=v\mu_1$, we find from (8.5) the equation

$$P\frac{\partial(v\mu_1)}{\partial y}-Q\frac{\partial(v\mu_1)}{\partial x}+v\mu_1\left\{\frac{\partial P}{\partial y}-\frac{\partial Q}{\partial x}\right\}=0$$

which, on account of (8.4), reduces to

$$\mu_1 \left\{ P\frac{\partial v}{\partial y} - Q\frac{\partial v}{\partial x} \right\} = 0$$

so that $v = const.$ is any general integral of (8.1). But if $u = C$ is one form of this general integral, we have $v = F(u)$. Thus

$$\mu_2 = \mu_1 F(u),$$

where $F(u)$ is any arbitrary function of u.

Example 8.1 Referring to Ex. **7.1**, we see that solutions of that equation are

$$\frac{1}{x^2} \div \frac{1}{y^2} = const., \quad \frac{1}{x^2} \div \frac{1}{xy} = const., \quad \frac{1}{y^2} \div \frac{1}{xy} = const.,$$

all of which are equivalent to $y/x = C$. Furthermore, an infinite number of integrating factors is included in the formula

$$\mu = \frac{1}{xy} F\left(\frac{y}{x}\right).$$

§9 Special types of integrating factor

It may happen that an integrating factor can be found which depends on one variable only. Consider, for example, the circumstances in which

$$P + Qy' = 0$$

admits of an integrating factor $\mu(x)$ depending upon x alone. That being the case, (7.4) will become

$$Q\frac{\mathrm{d}\mu}{\mathrm{d}x} = \mu\left(\frac{\partial P}{\partial y} - \frac{\partial Q}{\partial x}\right)$$

or

$$\frac{1}{\mu}\frac{\mathrm{d}\mu}{\mathrm{d}x} = \left\{\frac{\partial P}{\partial y} - \frac{\partial Q}{\partial x}\right\} \Big/ Q. \tag{9.1}$$

Thus μ can be a function of x alone if the right-hand member of this equation is independent of y; μ is then determined by a quadrature. In other words the equation $P + Qy' = 0$ has an integrating factor depending on x alone if

$$\frac{1}{Q}\left(\frac{\partial P}{\partial y}-\frac{\partial Q}{\partial x}\right)$$

is a function of x alone.

Example 9.1

$$(1-xy)+(xy-x^2)y'=0.$$

In the above notation,

$$\left(\frac{\partial P}{\partial y}-\frac{\partial Q}{\partial x}\right)\Big/ Q=\frac{-x-(y-2x)}{xy-x^2}=-\frac{1}{x}.$$

Hence

$$\int\frac{\mathrm{d}\mu}{\mu}=-\int\frac{\mathrm{d}x}{x},\quad \log|\mu|=-\log|x| \quad \text{or} \quad \mu=\frac{1}{x}.$$

With this factor introduced, the equation becomes

$$\left(\frac{1}{x}-y\right)+(y-x)y'=0;$$

it is now exact and has the integral

$$\log|x|-xy+\tfrac{1}{2}y^2=C.$$

In the same way, assuming the existence of an integrating factor which is a function of $x+y$, say

$$\mu=f(x+y)\equiv f(z),$$

it will be found that

$$\frac{f'(z)}{f(z)}=-\left\{\frac{\partial P}{\partial y}-\frac{\partial P}{\partial x}\right\}\Big/(P-Q). \tag{9.2}$$

Therefore a necessary condition for the existence of such an integrating factor is that the right-hand member of (9.2) shall be a function of z, or $x+y$ along; μ is then determined by a quadrature.

Example 9.2

$$(5x^2+2xy+3y^3)+3(x^2+xy^2+2y^3)y'=0.$$

Condition (9.2) here becomes

$$\frac{f'(z)}{f(z)}=-\frac{6y^2-4x}{2x^2+2xy-3xy^2-3y^3}=\frac{2}{x+y}=\frac{2}{z}$$

and therefore the condition is satisfied. We have $f(z)=z^2$ and therefore the integrating factor is $(x+y)^2$. Introducing it and integrating, we obtain the general integral

$$(x^2+y^3)(x+y)^3=C.$$

Again, if it is assumed that

$$\mu=f(xy)=f(z)$$

the condition becomes

$$\frac{f'(z)}{f(z)}=-\left\{\frac{\partial P}{\partial y}-\frac{\partial}{x}\right\}\Big/(Px-Qy), \tag{9.3}$$

where the right-hand member must be a function of z, or xy, alone; here again μ is determined by a quadrature.

Example 9.3

$$(xy^3+2x^2y^2-y^2)+(x^2y^2+2x^3y-2x^2)y'=0.$$

Condition (9.3) becomes

$$\frac{f'(z)}{f(z)}=\frac{xy^2-2x^2y-2y+4x}{xy^2-2x^2y}=1-\frac{2}{xy}=1-\frac{2}{z}$$

whence $f(z)=e^z z^{-2}$. An integrating factor is therefore $\mu=e^{xy}x^{-2}y^{-2}$; it leads to the general integral

$$e^{xy}\left(\frac{1}{x}+\frac{2}{y}\right)=C.$$

Similarly, it can be shown that a necessary condition for $\mu=f(y/x)$ to be an integrating factor of $P+Qy'=0$ is that

$$\left(\frac{\partial Q}{\partial x}-\frac{\partial P}{\partial y}\right)(P/y+Q/x)^{-1}$$

is a function of y/x alone. Denoting this function by $g(y/x)$, we find that the equation for the determination of f is

$$zf'(z)=f(z)g(z).$$

§10 The linear equation

An equation of the type

$$f(x)y' + g(x)y = h(x),$$

which involves y and y' linearly, is said to be *linear*. By dividing throughout by $f(x)$ it may be brought into the standard form

$$y' + py = q, \tag{10.1}$$

where p and q are functions of x alone. When condition (9.1) is applied to this equation it reveals the existence of an integrating factor which is a function of x alone. In fact, if the factor $\mu(x)$ is introduced the condition of integrability reduces to

$$\mu'(x) = \mu p \tag{10.2}$$

Integrating this equation we have

$$\log \mu = \int p \, dx,$$

so that a suitable integrating factor is

$$\mu(x) = \exp\left(\int p(x) \, dx\right).$$

The exact equation

$$e^{\int p \, dx} y' + (py - q) e^{\int p \, dx} = 0$$

has the general integral

$$y e^{\int p \, dx} - \int q e^{\int p \, dx} dx = C, \tag{10.3}$$

which thus involves two quadratures.

Note 1. When the integrating factor μ has been obtained and introduced, the equation may be written

$$\mu y' + \mu' y = q$$

with $\mu p = \mu'$, or

$$\frac{d}{dx}(\mu y) = q\mu.$$

Integrating both sides of this equation we obtain the equation

$$\mu y = \int q\mu \, dx + C.$$

The general integral is therefore of the form

$$y = C\mu^{-1} + \mu^{-1} \int q\mu \, dx. \tag{10.4}$$

Note 2. Replacing q by zero in (10.1) we obtain the corresponding reduced equation, homogeneous in y and y':

$$y'+py=0; \tag{10.5}$$

its general integral involves only one quadrature:

$$y=C\,e^{-\int p\,dx}.$$

Note 3. If any particular solution of (10.1), say $y=y_1$, is known, the general integral is obtainable by a quadrature. For we then have

$$y_1'+py_1=q,$$

and eliminating q between this equation and (10.1), also writing v for $y-y_1$, we have

$$v'+pv=0,$$

i.e. the corresponding reduced equation. Thus the general integral of (10.1) is

$$\begin{aligned} y&=v+y_1\\ &=C\,e^{-\int p\,dx}+y_1. \end{aligned} \tag{10.6}$$

It consists of two terms: the first is taken from the general integral of the reduced equation and is known as the *complementary function*, the second is any *particular integral*.

Note 4. The difference between any two particular integrals is a special case of the complementary function. For since the relation

$$y-y_1=C\,e^{-\int p\,dx}$$

is general, any particular integral y_2 is obtained by attributing the appropriate value to C, say C':

$$y_2-y_1=C'\,e^{-\int p\,dx}.$$

Note 5. If y_1 and y_2 are any two particular integrals,

$$\frac{y-y_1}{y_2-y_1}=const. \tag{10.7}$$

furnishes the general integral.

Example 10.1

$$xy'-(x+1)y=x^2-x^3.$$

Dividing by the coefficient of y' we reduce the equation to the standard form

$$y' - \frac{x+1}{x}y = x - x^2.$$

If μ is the integrating factor,

$$\log \mu = -\int \frac{x+1}{x}\,\mathrm{d}x = -x - \log x$$

or

$$\mu = \mathrm{e}^{-x}/x.$$

Introducing μ into the equation in its standard (not its original) form, we have

$$\frac{\mathrm{e}^{-x}y'}{x} - \frac{x+1}{x^2}\mathrm{e}^{-x}y = \mathrm{e}^{-x}(1-x)$$

or

$$\frac{\mathrm{d}}{\mathrm{d}x}\left(\frac{\mathrm{e}^{-x}y}{x}\right) = \mathrm{e}^{-x}(1-x),$$

so that the general integral is

$$y = Cx\,\mathrm{e}^{x} + x^2.$$

Application. A condenser of capacity C (farads) is being charged from a source of electricity of potential E (volts) through a non-inductive resistance R (ohms). The charge Q (coulombs) at time t (seconds) is given by the linear differential equation

$$R\frac{\mathrm{d}Q}{\mathrm{d}t} + \frac{Q}{C} = E.$$

If the initial charge is zero, it is required to find Q at any instant (i) when the voltage E is constant; (ii) when the voltage is alternating, i.e. $E = E_0 \sin pt$, where E_0 and p are constants.

Following the general method of integration we find that, whatever E may be,

$$Q\,\mathrm{e}^{t/RC} = K + \int \frac{E}{R}\,\mathrm{e}^{t/RC}\,\mathrm{d}t,$$

where K is the constant of integration.

(i) When E is constant,

$$Q\,e^{t/RC} = K + CE\,e^{t/RC}.$$

The constant K is determined by the initial condition that $Q=0$ when $t=0$, so that

$$0 = K + CE,$$

i.e.

$$Q\,e^{t/RC} = CE(e^{t/RC} - 1).$$

Thus the charge at time t is given by

$$Q = CE(1 - e^{-t/RC}),$$

which approximates to the steady value CE when t is large.

(ii) When $E = E_0 \sin pt$

$$e^{t/RC} Q = K + \frac{E_0}{R} \int e^{t/RC} \sin pt\, dt$$

$$= K + CE_0\, e^{t/RC}(\sin pt - RCp \cos pt)/(1 + R^2C^2p^2).$$

Since $Q=0$ when $t=0$ we have

$$0 = K - RC^2pE_0/(1 + R^2C^2p^2)$$

and hence

$$Q = \frac{CE_0}{1 + R^2C^2p^2}(\sin pt - RCp \cos pt + \mathrm{RCp}\, e^{-t/RC}).$$

The influence of the exponential term diminishes rapidly, so that Q approximates to a form of the same period as E.

§11 The Bernoulli equation

The equation

$$y' + py = qy^n$$

in which p and q are functions of x alone, is associated with the name of James Bernoulli. It may be reduced to linear form by a change in the dependent variable. For the result of division throughout by y^n:

$$y^{-n}y' + py^{1-n} = q$$

suggests the substitution

$$v=y^{1-n}, \qquad v'=(1-n)y^{-n}y'.$$

The equation then becomes

$$v'+(1-n)pv=(1-n)q, \tag{11.2}$$

which is of the standard linear type in v.

Example 11.1

$$\frac{\mathrm{d}y}{\mathrm{d}x}-\frac{y}{2x}=5x^2y^5.$$

The above steps give, in turn:

$$\frac{y'}{y^5}-\frac{1}{2xy^4}=5x^2$$

$$v=y^{-4}, \quad v'=-4y^{-5}y'$$

$$v'+\frac{2}{x}v=-20x^3.$$

This linear equation has the integrating factor x^2:

$$x^2v'+2xv=-20x^4.$$

Integrating:

$$x^2v=C-4x^5 \qquad \text{or} \qquad v=Cx^{-2}-4x^3$$

and since $y=v^{-1/4}$,

$$y=\frac{1}{\sqrt[4]{(Cx^{-2}-4x^3)}}.$$

§12 The Riccati equation

The standard form of this equation is

$$y'=py^2+qy+r \tag{12.1}$$

where p, q, r are functions of x alone, and p is not identically zero. It may be integrated completely when any particular solution, say $y=y_1$, is known, a result which is achieved by the substitution

$$y=y_1+1/v,$$

where v is a new dependent variable. After simplification it will be found that the equation in v is linear, namely:

$$v'+(2py_1+q)v+p=0 \tag{12.2}$$

and the integration can be completed by quadratures.

Note 1. The general integral may be expressed directly in terms of any three functions y_1, y_2, y_3 which satisfy the equation. For if we make the substitution $y=y_1+1/v$, y_2 and y_3 will correspond to two particular values of v, say v_1 and v_2, thus:

$$y_2=y_1+1/v_1, \qquad y_3=y_1+1/v_2,$$

and since $v=v_1$ and $v=v_2$ are particular solutions of the linear equation (12.2) its general solution will be (10.7)

$$\frac{v-v_1}{v_2-v_1}=C.$$

Making the reverse substitutions

$$v=\frac{1}{y-y_1}, \qquad v_1=\frac{1}{y_2-y_1}, \qquad v_2=\frac{1}{y_3-y_1}$$

we find that

$$\frac{y-y_2}{y-y_1}=C\frac{y_3-y_2}{y_3-y_1}$$

which furnishes the general integral.

Note 2. If $y=y_4$ is a fourth particular solution, we have

$$\frac{y_4-y_2}{y_4-y_1}\cdot\frac{y_3-y_1}{y_3-y_2}=C,$$

i.e. the cross-ratio of any four solutions is a constant.

Example 12.1

$$2x^2y'=(x-1)(y^2-x^2)+2xy.$$

The solutions $y=x$, $y=-x$ are evident on inspection. Taking the former, we write

$$y=x+1/v, \qquad y'=1-v'/v^2$$

and derive the linear equation

$$2x^2(v'+v)=1-x;$$

its general integral is

$$v=(Cx\,e^{-x}-1)/2x.$$

The general integral of the proposed equation is therefore

$$y=x+\frac{2x}{Cx\,e^{-x}-1}=x\frac{Cx\,e^{-x}+1}{Cx\,e^{-x}-1}=x\frac{Cx+e^x}{Cx-e^x}.$$

The cross-ratio of the four particular functions x, $-x$, $[x(x+e^x)/(x-e^x)]$, $[x(x-e^x)/(x+e^x)]$ which satisfy the equation will be found to be -1.

§13 Change of variable

To sum up, equations which may be readily integrated by quadratures are those with separable variables, those of the homogeneous type, exact and linear equations. In the homogeneous type, the variables are made separable by the substitution $v=y/x$ by which a change of dependent variable is effected. Actually the method of integrating an equation with separable variables, or a linear equation, is to introduce an integrating factor which renders it exact; this is the fundamental principle upon which integration of equations of the first order and degree essentially depends. If other equations can be integrated it is because a change in variable (as in the case of the Bernoulli equation) has enabled a reduction to one of the above types to be performed.

When an equation is proposed for solution and does not appear to come under one or other of the foregoing headings, it is advisable to consider if a change in one or both of the variables cannot be discovered which will transform the equation into one of a recognizable type, just as a change in the dependent variable of a Bernoulli equation rendered it linear.

For instance, if an equation is of the form

$$y'(x)=f(ax+by+c),$$

the substitution $v=ax+by$ at once suggests itself. It transforms the equation into

$$v'(x)=a+bf(v+c),$$

where the variables are separable.

Again, an equation such as

$$\{xf(y)+g(y)\}\frac{\mathrm{d}y}{\mathrm{d}x}=h(y)$$

may not be integrable as it stands, but if y is taken as the independent and x as the dependent variable, it may be written

$$h(y)\frac{\mathrm{d}x}{\mathrm{d}y}-f(y)x=g(y),$$

when it becomes a linear equation in x.

It is not possible, even if it were desirable, to formulate a set of rules governing changes of variable, but one or two combinations of symbols frequently occur and deserve a special mention. Thus if the differential $x+yy'$ is a component part of the equation, it naturally suggests $u=x^2+y^2$ as a new variable; if $xy'-y$ occurs it suggests $v=x/y$ or $v=y/x$. When both $x+y'$ and $xy'-y$ are present, simplification is very often obtained by a change to polar coordinates through the substitutions

$$x=r\cos\theta, \qquad y=r\sin\theta,$$

$$x\,\mathrm{d}x+y\,\mathrm{d}y=r\,\mathrm{d}r, \qquad x\,\mathrm{d}y-y\,\mathrm{d}x=r^2\,\mathrm{d}\theta.$$

In this case it is usually best to leave the general integral expressed in terms of r, θ.

Example 13.1

$$(xy+1)+2x^2(2xy-1)y'=0.$$

The prominence of xy in the equation suggests that $v=xy$ might be taken as a new variable in place of y. The equation then becomes

$$(4v^2-3v-1)-2x(2v-1)v'=0.$$

The variables are separable; the usual process of integration leads to the general integral

$$(4xy+1)^3(xy-1)^2=Cx^5.$$

Example 13.2 An equation of the type

$$xy'-y=f(x)g(y/x)$$

is transformed by the substitution $y=vx$ into

$$x^2v'=f(x)g(v),$$

whose variables are separable. For instance,

$$xy' - y = 2x\frac{y^2 - x^2}{x^4 - 1}$$

becomes, after transformation and separation

$$\frac{v'}{v^2 - 1} = \frac{2x}{x^4 - 1}$$

whence

$$\frac{v-1}{v+1} = C\frac{x^2 - 1}{x^2 + 1}.$$

If $c = (1 - C)/(1 + C)$, the general integral is

$$y = x\frac{x^2 + c}{cx^2 + 1}.$$

Example 13.3

$$(x^2 + y^2)(x + yy') + (x^2 + y^2 - 2x + 2y)(y - xy') = 0.$$

Changing to polar coordinates, we obtain, cancelling r^3,

$$\mathrm{d}r/\mathrm{d}\theta - (r - 2\cos\theta + 2\sin\theta) = 0.$$

This is linear in r and has the general integral

$$r = C\,\mathrm{e}^{\theta} - 2\sin\theta.$$

2

Integral curves

§14 Families of plane curves

When referred to a set of rectangular axes of coordinates, the equation $f(x, y)=c$ represents, for any definite value of the parameter c, a plane curve. The aggregate of all values of c for which the equation is possible gives a complete one-parameter family of curves. Thus the equation $x^2+y^2=c$, in which all positive values are in turn assigned to c, represents the family of all circles whose centres are at the origin of coordinates.

The corresponding differential equation

$$P(x, y)+Q(x, y)y'=0, \tag{14.1}$$

in which $P=\partial f/\partial x$ and $Q=\partial f/\partial y$, does not involve the parameter c; it is therefore representative of some geometrical property or other that is common to all members of the family of curves. Any pair of values (x, y) for which $P(x, y)$ and $Q(x, y)$ have definite real values correspond to a point in the plane at which the value of the gradient $y'=\mathrm{d}y/\mathrm{d}x$ can be evaluated by means of equation (14.1). The only exception to this statement arises when the ratio P/Q becomes indeterminate, a case that will be investigated later; see §17.

Thus the integration of the linear differential equation (14.1) is equivalent to the geometrical problem of discovering the family of plane curves, called the *integral curves*, such that the gradient at any point (x, y) is the assigned function $-P(x, y)/Q(x, y)$ of the pair of variables (x, y). If the functions P and Q are single-valued, there will be one, and only one, value of y' corresponding to each point for which the ratio P/Q is determinate. This amounts to the statement that through any point (x, y) there passes, in general, only one integral curve of a first-order linear differential equation.

The symbolic expression of any geometrical property that involves, directly or indirectly, the gradient at any point is in fact a differential equation of the first order. It may, however, be of higher order than the

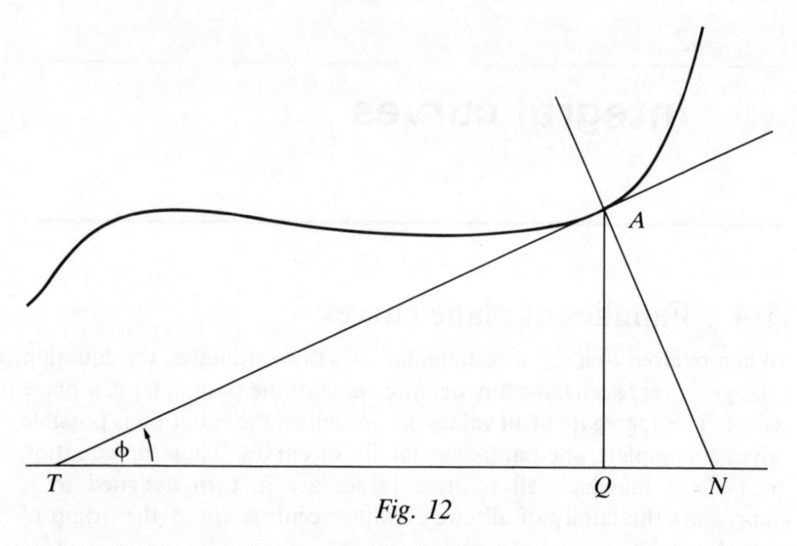

Fig. 12

first, but for the moment we shall confine our attention to problems that lead to equations of the first order.

Suppose that A (see Fig. 12) is the point with coordinates (x, y) on a plane curve, and that T, Q and N are the points at which the tangent, ordinate and normal at A respectively meet the x-axis. Then if ϕ is the angle between TA and the positive direction of the x-axis, $y' = \tan\phi$. The gradient may be introduced indirectly through the lengths AN, QN and TQ. The length AN, known as the *normal*, is given by

$$AN = y \operatorname{cosec}\phi = y(1 + y'^2)^{1/2}/y'$$

so it does not involve y' linearly. The *subnormal*, QN, is given by

$$QN = y \tan\phi = yy',$$

and TQ, known as the *subtangent*, by

$$TQ = y \cot\phi = y/y'.$$

The length AT is known as the *tangent*.

Example 14.1 Curves whose subnormal has the constant value $2a$ satisfy the linear differential equation $yy' = 2a$. Integrating this equation we obtain the general solution

$$y^2 = 4a(x + c)$$

which represents a family of equal parabolas, with axes coincident with the x-axis.

Example 14.2 Curves whose subtangent has the constant value a satisfy the differential equation $y/y' = a$. Integrating this equation we obtain the solution

$$\log y = \log c + x/a,$$

which is equivalent to

$$y = c\,e^{x/a},$$

representing a family of exponential curves.

Example 14.3 Curves in which the subtangent is twice the abscissa satisfy the differential equation $y/y' = 2x$. The variables are separable and then it is easily shown that the integral curves are the members of the family of parabolas $y^2 = cx$.

Example 14.4 Consider a curve in which the ordinate at any point is equal to the perpendicular from the origin to the normal at that point.

If (X, Y) is any point on the normal at (x, y), then

$$X + Yy' = x + yy',$$

and therefore the length of the perpendicular from the origin is

$$(x + yy')(1 + y'^2)^{-1/2},$$

so that curves with the stated property satisfy the differential equation

$$x + yy' = y(1 + y'^2)^{1/2}.$$

In that form, the equation does not appear to be linear, but on rationalization and simplification it reduces to the linear form

$$x^2 + 2xyy' = y^2.$$

This can be integrated as an equation of *homogeneous type* (cf. Example 4.1) or as one which has an integrating factor which is a function of x alone. The second method is the simpler; the integrating factor x^{-2} leads to the general integral

$$x^2 + y^2 = cx,$$

representing a family of circles touching the y-axis at the origin.

§15 Trajectories

Suppose that

$$f(x, y, c) = 0, \qquad g(x, y, k) = 0,$$

are the equations of two families of curves each dependent on a single parameter. When each member of the k-family cuts every member of the c-family according to a definite law, any curve of either family is said to be a *trajectory* of the other family. The most important case is that in which the curves of the two families intersect at a constant angle.

When a curve intersects all the members of a family at right angles, it is said to be an *orthogonal trajectory* of that family. Thus any circle of centre 0 is the orthogonal trajectory of the family of straight lines radiating from 0, and conversely any line through 0 is the orthogonal trajectory of all circles concentric with 0. In other words, the families

$$x^2+y^2=c^2, \qquad y=kx$$

are mutually orthogonal

In two-dimensional electrostatics, the equipotential lines are the orthogonal trajectories of the lines of force.

Suppose that, in rectangular cartesian coordinates, the linear differential equation describing one family of curves is

$$P(x, y)+Q(x, y)y'=0; \tag{15.1}$$

when written in the form

$$y'=-P(x, y)/Q(x, y),$$

this equation expresses the fact that if an integral curves passes through the point (x, y) its tangent at that point will have gradient $-P(x, y)/Q(x, y)$. The gradient of the orthogonal curve through that point will be the negative reciprocal, i.e. the tangent at (x, y) to the orthogonal curve will have gradient

$$y'=Q(x, y)/P(x, y).$$

In other words, an orthogonal trajectory of the system of curves described by equation (15.1) will be an integral curve of the linear differential equation

$$P(x, y)y'-Q(x, y)=0. \tag{15.2}$$

Conversely, whenever an integral curve of the differential equation (15.2) meets an integral curve of the differential equation (15.1), it will cut it at right angles.

Thus the differential equation of one family is obtained from that of the other by replacing y' by $-1/y'$ and we have the following result: *If the differential equation of the one-parameter family of curves $f(x, y)=c$ is $F(x, y, y')=0$, then the differential equation describing the family of* *orthogonal trajectories is* $F(x, y, -1/y')=0$, *and its general integral* $g(x, y, k)=0$ *will be the equation of the family of orthogonal trajectories.*

When the curves of the families $f(x, y, c)=0$ and $g(x, y, k)=0$ intersect at a constant angle ω, other than a right angle, the trajectories are said to be *oblique*. To make the problem precise, suppose that the tangent at the point (x, y) to a curve of the c-family makes an angle ϕ_1 with the x-axis and that the inclination of the tangent to the k-curve at (x, y) is ϕ_2; then

$$\tan \phi_1 = \frac{\tan \phi_2 - m}{1 + m \tan \phi_2},$$

where $m = \tan \omega$.

Suppose that the differential equation of the original family is $F(x, y, y')=0$, then this may be written as $F(x, y, \tan \phi_1)=0$, or again as

$$F\left(x, y, \frac{\tan \phi_2 - m}{1 + m \tan \phi_2}\right)=0.$$

However, in the family of oblique trajectories $\tan \phi_2 = y'$ and so this family consists of the integral curves of the differential equation

$$F\left(x, y, \frac{y' - m}{1 + my'}\right)=0.$$

If therefore the differential equation of a family of plane curves is known, the differential equation of the family of oblique trajectories that intersects it at a positive angle ω is obtained by replacing y' whever it occurs by $(y'-m)/(1+my')$, where $m = \tan \omega$.

In many problems it is more convenient to use polar coordinates (r, θ) rather than cartesian coordinates. Suppose that $f(r, \theta, c)=0$ represents a one-parameter family of plane curves in polar coordinates, and that $F(r, \theta, \dot{\theta})$, where $\dot{\theta} = d\theta/dr$, is the corresponding differential equation. If (r, θ) is any point through which a curve of the family passes and ϕ_1 is the angle which the tangent to the curve at that point makes with the positive direction of the radius vector, then

$$\tan \phi_1 = r\dot{\theta}.$$

An oblique trajectory, of angle ω, is such that the angle its tangent makes with the radius vector is $\phi_2 = \phi_1 + \omega$, and we have

$$\tan \phi_1 = \frac{\tan \phi_2 - \tan \omega}{1 + \tan \phi_2 \tan \omega} = \frac{r\dot{\theta} - m}{1 + mr\dot{\theta}},$$

where, as before $m = \tan \omega$. Thus the differential equation of the oblique trajectories will be obtained from that of the original family by replacing $r\dot{\theta}$ by $(r\dot{\theta}-m)/(1+mr\dot{\theta})$.

In the case of orthogonal trajectories we should have $\phi_2=\phi_1+\frac{1}{2}\pi$ or $\tan\phi_2=-\cot\phi_1=-1/r\dot{\theta}$, so that $r\dot{\theta}$ is replaced by $-1/r\dot{\theta}$.

Example 15.1 The equation $x^2+y^2=2cx$ represents the system of circles touching the y-axis at the origin, the centre of any individual circle being $(c, 0)$. The differential equation of the family is (cf. Example 14.4)

$$x^2-y^2+2xyy'=0.$$

To find the equation of the orthogonal trajectories we replace y' by $-1/y'$ to obtain

$$(x^2-y^2)y'-2xy=0.$$

This is a homogeneous equation which we could solve by changing the independent variable to $v=y/x$; however, if we write the equation in the form

$$y^2-x^2+2yx\frac{dx}{dy}=0,$$

we see that it is equivalent to the differential equation of the original family with x and y interchanged throughout. Its general integral is therefore

$$x^2+y^2=2ky,$$

so that the orthogonal trajectories are all circles which touch the z-axis at the origin.

Example 15.2 It may happen that any one member of a one-parameter family of curves intersects at right angles an infinite set (but not necessarily all) of members of the same set. Such a family is said to be *self-orthogonal.* The differential equation of the family must therefore remain invariant under the substitution of $-1/y'$ for y'. Obviously this cannot happen if the differential equation of the family is linear.

For example, the differential equation

$$y(y'^2-1)+2xy'=0$$

of the family of parabolas

$$y^2=4c(x+c)$$

is unaltered if y' is replaced by $-1/y'$. The original family is therefore self-

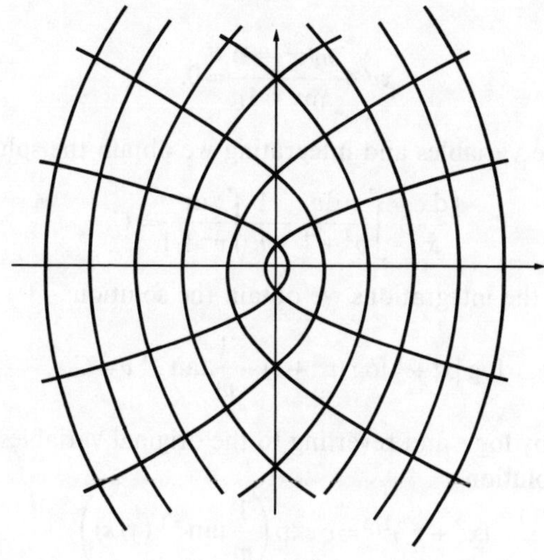

Fig. 13

orthogonal (cf. Fig. 13). To scc how this property originates, consider the two particular curves

$$y^2 = 4a(x+a), \qquad y^2 = 4b(x+b).$$

Their point of intersection has the abscissa $x = -(a+b)$ and the ordinate is given by $y^2 = -4ab$. Thus the intersection is real only when a and b are of opposite signs. When that is the case, the gradients of the two curves at the point of intersection are respectively either $(-a/b)$ and $-(-a/b)$ or $-(-a/b)$ and $(-b/a)$, which verifies their orthogonality.

Example 15.3 To find the oblique trajectories of the family of straight lines $y = cx$ we note that the differential equation describing this family is $y' = y/x$ and therefore that the differential equation of the oblique trajectories is

$$x(y' - m) = y(m + y') \tag{15.3}$$

or

$$y' = -\frac{mx + y}{my - x}$$

This equation is homogeneous so we put $y = xv$ to obtain the equation

$$xv' + v + \frac{m + v}{mv - 1} = 0$$

or

$$xv' + \frac{m(v^2+1)}{\text{mv}-1)} = 0.$$

Separating the variables and integrating we obtain the solution

$$\int \frac{\mathrm{d}x}{x} + \int \frac{v\,\mathrm{d}v}{v^2+1} - \frac{1}{m}\int \frac{\mathrm{d}v}{v^2+1} = C.$$

Carrying out the integrations we obtain the solution

$$\log|x| + \tfrac{1}{2}\log(v^2+1) - \frac{1}{m}\tan^{-1} v = C.$$

Replacing C by $\log c$ and reverting to the original variables, we obtain the general solution

$$(x^2+y^2)^{1/2} = c\exp\left(\frac{1}{m}\tan^{-1}(y/x)\right)$$

which, in polar coordinates, takes the simple form

$$r = c\,\mathrm{e}^{\theta/m} \tag{15.4}$$

i.e. the oblique trajectories are a family of equiangular spirals.

The form of solution (15.4) suggests that it would have been easier to use polar coordinates from the start. The family of straight lines then has equation $\theta = \alpha$ and the corresponding differential equation is $\dot{\theta} = 0$. The differential equation of the oblique trajectories is therefore

$$\frac{r\dot{\theta} - m}{1 - mr\dot{\theta}} = 0$$

or simply

$$r\frac{\mathrm{d}\theta}{\mathrm{d}r} = m.$$

Again the variables are separable and we obtain the solution (15.4)

Example 15.4 In polar coordinates the family of circles considered in Example 15.1 has equation

$$r = 2c\cos\theta,$$

and its differential equation is

$$\frac{\mathrm{d}\theta}{\mathrm{d}r} = -\frac{\cot\theta}{r}.$$

The orthogonal trajectories, which are the integral curves of

$$r\frac{d\theta}{dr}=\tan\theta,$$

are the family

$$r=2k\sin\theta.$$

§16 Level lines and lines of slope of a surface

Suppose that, referred to a rectangular cartesian system of axes in which the z-axis is vertical and the xy-plane is the horizontal reference plane, a surface has equation $f(x, y, z)=0$. The section made on the surface by a horizontal plane $z=c$ is called a *level line* or *contour line* of level c. The horizontal projcctions of the various level lines made by a regular sequence of horizontal planes $z=c_1, c_2, \ldots$ form what is called a *contour map*. If c is varied continuously as it does when the secant plane moves uniformly upwards, the contour map is replaced by the family of curves $f(x, y, c)=0$.

The *lines of steepest slope* on the surface are the lines of minimum distance between successive contours, and are therefore lines cutting the level lines at right angles. On account of the fact that when two curves in space, one of which lines in a horizontal plane, cut orthogonally, their horizontal projections cut orthogonally, we see that on the horizontal reference plane, the projections of the lines of steepest slope are the orthogonal trajectories of the contour lines. If the equation of these trajectories is $g(x, y, k)=0$, the lines of steepest slope on the surface are the intersections of the surface by the cylinders $g(x, y, k)=0$ whose generators are parallel to the z-axis.

Example 16.1 Consider the paraboloid

$$2z=ax^2+by^2.$$

The level lines project into the family of curves

$$ax^2+by^2=2c,$$

whose differential equation is

$$ax+byy'=0.$$

The orthogonal trajectories, whose differential equation is

$$axy'-by=0$$

will be found to be $y^a = kx^b$, and hence the lines of steepest slope may be found.

In particular, when $a = b$, i.e. in the case of a paraboloid of revolution, the lines of steepest slope are the intersections of the surface by all planes through the z-axis; that is, they are meridian lines on the surface.

§17 Singular points

Suppose that (x_0, y_0) is a point such that, in its neighbourhood, the functions $P(x, y)$, $Q(x, y)$ are finite, continuous and single-valued. Then the equation

$$P(x, y) + Q(x, y)y' = 0 \tag{17.1}$$

assigns the value $y' = -P(x_0, y_0)/Q(x_0, y_0)$ to the gradient at that point, and this value is determinate with the sole exception of the case

$$P(x_0, y_0) = 0, \qquad Q(x_0, y_0) = 0. \tag{17.2}$$

Points at which the gradient becomes indeterminate through the simultaneous vanishing of P and Q are said to be *singular*. Thus singular points are the points of intersection of the curves $P(x, y) = 0$ and $Q(x, y) = 0$, which are the loci of points at which the integral curves have tangents parallel to the x-axis and the y-axis respectively. (It is assumed that $P = 0$ and $Q = 0$ have not a branch in common. If $R = 0$ were a branch common to both, R could be cancelled out of the equation, and the above statement would be strictly true.) They are thus, in general, isolated points.

To illustrate possible modes of behaviour of the integral curves in the neighbourhood of a singular point, we shall consider the linear fractional equation

$$y' = \frac{ax + by}{lx + my} \qquad (am - bl \neq 0) \tag{17.3}$$

for which the origin is the only finite singular point.

Making the usual substitution $y = vx$, we obtain

$$xv' = \frac{a + (b - l)v - mv^2}{l + mv} \tag{17.4}$$

If $v - \alpha$, $v - \beta$ are factors of the numerator, then $v = \alpha$, $v = \beta$ are particular solutions of (17.4), and $y = \alpha x$, $y = \beta x$ of (17.3). Thus among the integral curves of (17.3) are two straight lines through the origin, whose joint equation is

$$ax^2+(b-l)xy-my^2=0.$$

These are known as the *principal lines*, and α, β the *principal directions* through the origin. They are real and distinct, coincident, or complex according as

$$(b-l)^2+4am>0, \quad =0 \quad \text{or} \quad <0.$$

When they are real, (17.4) may be written

$$xv'+\frac{m(v-\alpha)(v-\beta)}{l+mv}=0.$$

Separating the variables and integrating, we obtain

$$\log|x|+\mu\log|v-\alpha|+v\log|v-\beta|=\log C$$

or

$$x(v-\alpha)^{\mu}(v-\beta)^{v}=C,$$

where

$$\mu=\frac{l+m\alpha}{m(\alpha-\beta)}, \quad v=\frac{l+m\beta}{m(\beta-\alpha)}, \quad \mu+v=1,$$

and thus the general integral of (17.3), when the principal directions are real and distinct, is

$$(y-\alpha x)^{\mu}(y-\beta x)^{v}=C.$$

When μ and v are both positive, this represents a family of curves having asymptotes $y=\alpha x$, $y=\beta x$ in common. The symptotes themselves are integral curves, but apart from them, no integral curves pass through the origin, which is a *col* of the family. When μ and v are of opposite signs, the equation of the integral curves may be written

$$y-\beta x=c(y-x)^{\rho}, \qquad \text{where } \rho=-\mu/v.$$

Since ρ is positive, every integral curve passes through the origin, which is a *base point* or *nodal point* of the family. When $\rho=1$, i.e. $\mu=-v$, the integral curves pass through the origin in all directions; the nodal point is then said to be *isotropic*. When $\rho>1$, i.e. $|\mu|>|v|$, $y-\alpha x$ and $y=\beta x$ are integral curves; all other integral curves touch the principal line $y=\beta x$ at the origin. When $0<\rho<1$, i.e. $|\mu|<|v|$, $y=\alpha x$ and $y=\beta x$ are integral curves; all other integral curves touch $y=\alpha x$ at the origin.

When the principal directions coincide with that of the line $y=\alpha x$,

(17.4) may be written

$$xv' + \frac{m(v-\alpha)^2}{l+mv} = 0,$$

whence

$$\log|x| + \log|v-\alpha| - \frac{l+m\alpha}{m(v-\alpha)} = C,$$

and thus the general integral of (17.3) is

$$(y-\alpha x)(\log|y-\alpha x| - C) = (l/m+\alpha)x,$$

from which we derive

$$y' = \alpha + \frac{l/m+\alpha}{l - C + \log|y-\alpha x|}.$$

Thus the line $y=\alpha x$, which is itself an integral curve, is tangential at the origin to every other integral curve.

When the principal directions are complex, we may write $\alpha=\kappa+i\lambda$, $\beta=\kappa-i\lambda$, whereupon (17.4) becomes

$$xv' + \frac{m\{(v-\kappa)^2+\lambda^2\}}{l+mv} = 0.$$

Separating the variables and integrating, we have

$$\log|x| + \tfrac{1}{2}\log\{(v-\kappa)^2+\lambda^2\} + \frac{l+m\kappa}{m\lambda}\arctan\frac{v-\kappa}{\lambda} = \log C,$$

and hence the general integral of (17.3) is

$$\tfrac{1}{2}\log\{(y-\kappa x)^2+\lambda^2 x^2\} + \frac{l+m\kappa}{m\lambda}\arctan\frac{y-\kappa x}{\lambda x} = \log C.$$

If $l+m\kappa=0$, the integral curves

$$(y-\kappa x)^2+\lambda^2 x^2 = C^2$$

are a family of ellipses encircling the singular point at the origin, which is their centre and limiting point. In the general case we make the substitution†

$$y-\kappa x = r\sin\theta, \qquad \lambda x = r\cos\theta$$

† This transformation may be regarded as a strain $\eta=y-\kappa x$, $\xi=\lambda x$ consisting of a shear in the direction of the y-axis and an extension in the direction of the x-axis, followed by a change to polar coordinates.

and obtain

$$\log r + \mu\theta = \log C, \qquad \text{where } \mu = (l + m\kappa)/m\lambda,$$

or

$$r = C\,\mathrm{e}^{-\mu\theta}.$$

This is the equation of an equiangular spiral which winds asymptotically round the origin. The transformation back to the xy-plane involves no deformation other than a simple strain, so that the integral curves are in general spirals with a common asymptotic point at the origin.

Note. When (17.1) has a singular point at the origin it may be written

$$\frac{\mathrm{d}y}{\mathrm{d}x} = \frac{ax + by + p(x, y)}{lx + my + q(x, y)},$$

where $p(x, y)$, $q(x, y)$ can be developed as series whose lowest terms are of the second degree at least. It can be shown that unless $a = b = 0$ or $l = m = 0$ the integral curves behave, in the neighbourhood of the origin, as if p and q were absent, and precisely the same cases arise.

Example 17.1

$$\frac{\mathrm{d}y}{\mathrm{d}x} = \frac{my - x^n}{x} \qquad (n \neq m,\ n > 1).$$

The general integral $y = Cx^m - x^n/(n - m)$ shows that y is of the order of x^m or x^n according as $m <$ or $> n$. When $m < 0$, the origin is a *col* of the integral curves; when $m > 0$, it is a *nodal point*, with integral curves touching the y-axis when $m < 1$, and the x-axis when $m > 1$.

Example 17.2

$$\frac{\mathrm{d}y}{\mathrm{d}x} = -\frac{x + 2x^3}{y}.$$

The integral curves $y^2 + x^2 + x^4 = C^2$ are closed ovals; the origin is their limiting point or centre.

Example 17.3

$$\frac{\mathrm{d}y}{\mathrm{d}x} = \frac{-x + \mu y - \mu y(x^2 + y^2)}{y + \mu x - \mu x(x^2 + y^2)}.$$

When transformed into polar coordinates, this equation becomes

$$\mu \frac{d\theta}{dr} = \frac{1}{r(r^2 - 1)}$$

and the general integral is $r^2(1 + C\,e^{2\mu\theta}) = 1$. The integral curves are spirals with the origin as *asymptotic point.*

3

Equations of higher degree

§18 The general integral

When the primitive is of the form $f(x, y, C)=0$, involving the arbitrary constant C implicitly, the differential equation arises through elimination of C between the equation

$$f=0, \quad \frac{\partial f}{\partial x}+\frac{\partial f}{\partial y}y'=0.$$

Only in exceptional cases does the result of this elimination involve y' to the first degree; if the resulting equation

$$F(x, y, y')=0 \tag{18.1}$$

is a polynomial in y', involving y' to the mth power, it is said to be of *degree m.* But it may be irrational in y' or even transcendental.

Theoretically, (18.1) may be solved for y', giving a set of equations

$$y'=F_1(x, y), \qquad y'=F_2(x, y), \tag{18.2}$$

each of the form hitherto considered. This set may be finite or, in the case of a transcendental equation, infinite in number. If the general integrals of the equations (18.2) are respectively

$$f_1(x, y, C)=0, \qquad f_2(x, y, C)=0, \tag{18.3}$$

the *general integral* of (18.1) is defined to be any equation $f(x, y, C)=0$ which is satisfied when, and only when, at least one equation (18.3) is satisfied. In particular, when the equation is of degree m, its general integral is the product of the m integrals (18.3), that is

$$f_1(x, y, C)f_2(x, y, C)\cdots f_m(x, y, C)=0.$$

In practice, however, the decomposition of (18.1) into (18.2). and the solution of these latter equations, or both, may present difficulties, in which case other lines of approach, some of which will be indicated in the following sections, may be attempted.

Errors in writing and printing are apt to arise through confusion between y' and y; these are obviated by writing p for y', a convention that will be adhered to throughout the present chapter.

Example 18.1 Let the primitive be the equation of a one-parameter family of straight lines; if the gradient m of each straight line be taken to serve as its parameter, the equation may be written

$$y = mx + f(m),$$

where f is some specified function. Then

$$p = y' = m.$$

Eliminating m, we obtain the differential equation

$$y = px + f(p).$$

In particular, the equation of the family $y = mx + c/m$, where c is a constant, is

$$y = px + cp$$

or

$$p^2 x - py + c = 0,$$

of the second degree.

Example 18.2 The second degree equation

$$p^2 - 2px + x^2 - y^2 = 0,$$

may be decomposed into the pair

$$p - y - x = 0, \qquad p + y - x = 0.$$

These are linear equations whose general integrals may be written

$$y - C\,\mathrm{e}^{x} + x + 1 = 0, \qquad y - C\,\mathrm{e}^{-x} - x + 1 = 0.$$

Multiplying these two integrals together, we obtain

$$C^2 - C\{\mathrm{e}^{x}(y - x + 1) + \mathrm{e}^{-x}(y + x + 1)\} + (y + 1)^2 - x^2 = 0,$$

as the general solution of the given equation.

§19 The Clairaut equation

We consider the equation obtained in the last section (Ex. **18.1**)

$$y = px + f(p), \tag{19.1}$$

known as the *Clairaut equation*, with a view to discovering whether or not it admits of any integral other than the primitive from which it was derived. Differentiating with respect to x, we obtain

$$p=p+\{x+f'(p)\}\frac{\mathrm{d}p}{\mathrm{d}x}.$$

This equation may be satisfied in just two ways. On the one hand, we may take $\mathrm{d}p/\mathrm{d}x=0$, whence $p=C$, and thus recover the primitive

$$y=Cx+f(C). \tag{19.2}$$

On the other hand, the equation is satisfied if

$$x=-f'(p), \tag{19.3}$$

and if x is now eliminated from (19.1),

$$y=-pf'(p)+f(p). \tag{19.4}$$

Equations (19.3) and (19.4) taken together are the parametric equations of an integral curve of (19.1); by eliminating p we obtain the equation of this curve in a form $\phi(x, y)=0$ involving no arbitrary constant. This integral curve represents a singular solution (§2) of the equation. Since the envelope of the primitive family (19.2) is obtained by eliminating C between that equation and $x+f'(C)=0$, and since the result of this elimination is identical with that of eliminating p between (19.3) and (19.4), it follows that the singular solution may be interpreted geometrically as the envelope of the family of integral curves.

Example 19.1

$$y=px+a/p.$$

Differentiating,

$$p=p+(x-a/p^2)\,\mathrm{d}p/\mathrm{d}x.$$

Firstly, $\mathrm{d}p/\mathrm{d}x=0$ or $p=C$ gives the general integral

$$y=Cx+a/C.$$

Secondly, the alternative $x=a/p^2$ leads to $y=2a/p$, whence, eliminating p,

$$y^2=4ax.$$

Thus the singular solution represents the parabola enveloping the lines $y=Cx+a/C$, and hence the integral curves of the differential equation consist of the parabola together with the aggregate of its tangents. This fact may be confirmed by considering the equation in the form

$$p^2x - py + a = 0;$$

regarded as a quadratic in p, it has real roots only when $y^2 - 4ax \geqslant 0$, i.e. outside the parabola there are integral curves (the tangents), inside there are none. The parabola itself is an integral curve separating these two regions (cf. Fig. 14).

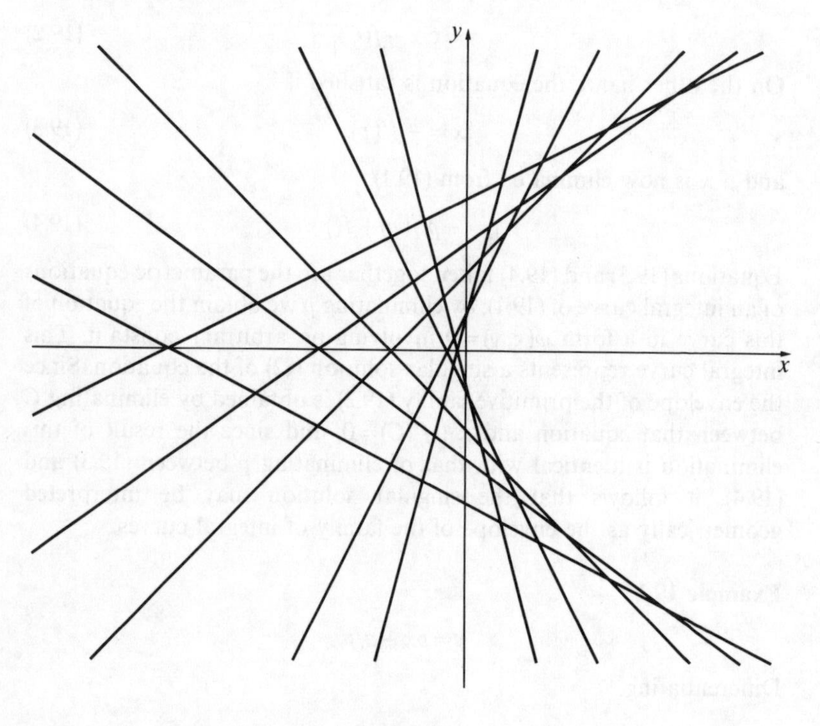

Fig. 14

Example 19.2 A geometrical problem sometimes leads to a differential equation whose singular integral furnishes the solution aimed at. For instance: *to find a curve such that the tangent at any point makes with the axes a triangle of constant area a^2*. Since the intercepts of the tangent at (x, y) on the x- and y-axes are $x\text{-}y/p$, $y - px$ respectively, the differential equation is

$$(x - y/p)(y - px) = 2a^2 \qquad \text{or} \qquad (y - px)^2 = -2a^2p.$$

This is equivalent to the Clairaut equation

$$y = px \pm a\sqrt{(-2p)},$$

whose singular solution, given parametrically by

$$x = \mp a\sqrt{(-1/2p)}, \qquad y = \pm \tfrac{1}{2}a\sqrt{(-2p)},$$

is $xy = \frac{1}{2}a^2$. Thus the curve sought is a rectangular hyperbola.

§20 Generalization – the d'Alembert equation

The Clairaut equation is a special case of the more general *d'Alembert equation* (also known as the *Lagrange equation*)

$$y = xg(p) + f(p); \tag{10.1}$$

in fact it is so exceptional a case that it will now be excluded by the stipulation that $g(p)$ is different from p. Differentiationg with respect to x, we have

$$p = g(p) + \{xg'(p) + f'(p)\}p'(x). \tag{20.2}$$

Since $g(p)$ does not cancel out with p, we cannot have $\mathrm{d}p/\mathrm{d}x = 0$ identically, and therefore the general integral curve is not a straight line. Nevertheless, the equation $g(p) - p = 0$ may have real roots; if $p = m$ is one such root $\mathrm{d}p/\mathrm{d}x$ will be zero, and (20.2) will be satisfied. Thus among the integrals of (20.1) there may be some of the linear form $y = xg(m) + f(m)$, where m is any one of the real roots of $g(p) - p = 0$. If will be shown later (§26, Note 2) that these are singular integrals, and not merely special integrals, of the equation.

Since $\mathrm{d}p/\mathrm{d}x$ is not identically zero, (20.2) may be divided throughout by it, and thus rewritten

$$\{g(p) - p\}\frac{\mathrm{d}x}{\mathrm{d}p} + g'(p)x + f'(p) = 0.$$

This is a linear equation for x; if

$$\log \phi(p) = \int \frac{\mathrm{d}p}{g(p) - p},$$

$\phi(p)$ is an integrating factor, and the general integral is

$$\{g(p) - p\}\phi(p)x = C - \textstyle\int f'(p)\phi(p)\,\mathrm{d}p. \tag{20.3}$$

Thus x is expressed in terms of p and the arbitrary constant C; by eliminating x between (20.1) and (20.3), y may be similarly expressed, and thus the general integral may be obtained in terms of p, regarded as a parameter. It is only exceptionally that the parameter can be eliminated and the general integral obtained as a single equation between x, y and C; usually (20.3) taken together with the equation itself must be considered to furnish the general integral.

Example 20.1

$$y=x+p^2-\tfrac{2}{3}p^3.$$

This is of the above form; differentiating with respect to x,

$$p=1+(2p-2p^2)p'$$

or

$$(p-1)(2pp'+1)=0.$$

The alternatives are therefore

$$p=1 \qquad \text{and} \qquad 2pp'+1=0.$$

Substituting $p=1$ in the original equation we find that $y=x+\frac{1}{3}$; this solution will be set aside for the moment.

From the second alternative we obtain

$$p^2+x=C,$$

and, deducing y from the original equation, we may express the general integral as

$$x=C-p^2, \qquad y=C-\tfrac{2}{3}p^3.$$

In this case p can be eliminated, to give the general integral in the form

$$9(C-y)^2=4(C-x)^3.$$

The line $y=x+\frac{1}{3}$ is not a member of this family of curves, but is their envelope. Consider in particular the intersections of this line with the particular curve ($C=0$)

$$9y^2+4x^3=0;$$

the abscissas are given by

$$(3x+1)^2+4x^3=0,$$

i.e.

$$4x^3+9x^2+6x+1=0 \qquad \text{or} \qquad (x+1)^2(4x+1)=0.$$

Thus the line touches the particular curve $C=0$ at a point of abscissa -1; since the other integral curves of the family are derived from this particular one by translation parallel to the line $y=x$ (which leaves the line $y=x+\frac{1}{3}$ invariant), every curve of the family touches the line $y=x+\frac{1}{3}$, which is their envelope (cf. Fig. 15).

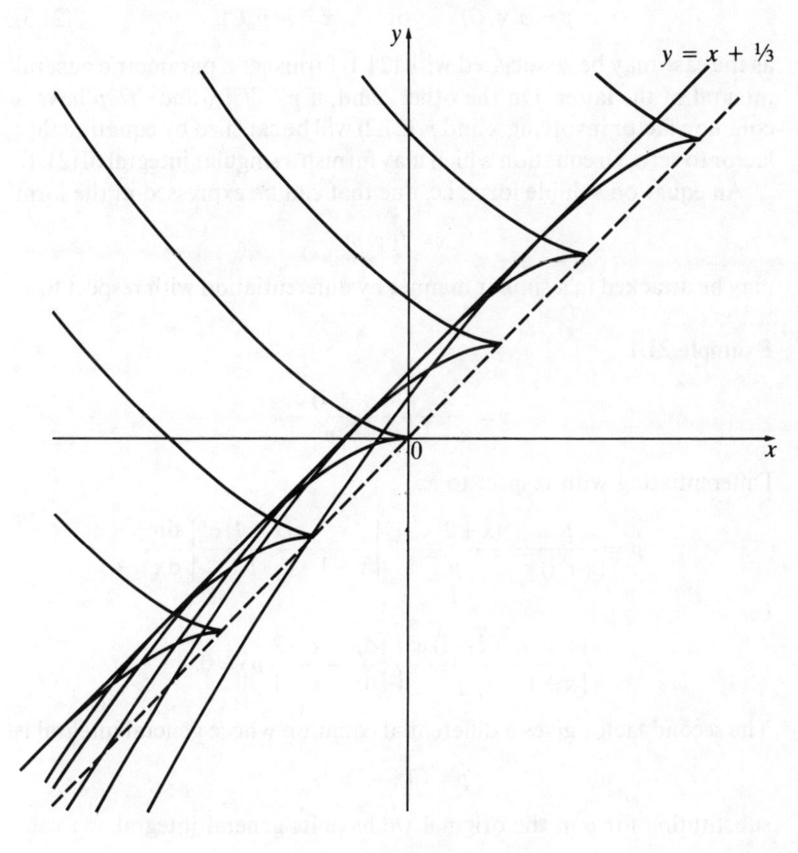

Fig. 15

§21 Further generalization

The integration of any differential equation that can be solved algebraically for y in terms of x and p, thus

$$y=f(x, p), \tag{21.1}$$

may be attempted by the same device of differentiating with respect to x. This gives

$$p=\frac{\partial f}{\partial x}+\frac{\partial f}{\partial p}\frac{\mathrm{d}p}{\mathrm{d}x}, \tag{21.2}$$

which is an equation of the first degree in $\mathrm{d}p/\mathrm{d}x$; its general integral

$$p=\phi(x, C) \qquad \text{or} \qquad x=\psi(p, C), \tag{21.3}$$

as the case may be, associated with (21.1) furnishes a parametric general integral of the latter. On the other hand, if $p-\partial f/\partial x$ and $\partial f/\partial p$ have a common factor involving x and p, (21.2) will be satisfied by equating that factor to zero, an equation which may furnish a singular integral of (21.1).

An equation soluble for x, i.e. one that can be expressed in the form

$$x=f(y, p),$$

may be attacked in a similar manner by differentiation with respect to y.

Example 21.1

$$y=\frac{x}{x+1}p+\frac{(x+1)\,\mathrm{e}^x}{p}.$$

Differentiating with respect to x:

$$p=\frac{p}{(x+1)^2}+\frac{(x+2)\,\mathrm{e}^x}{p}+\left\{\frac{x}{x+1}-\frac{(x+1)\,\mathrm{e}^x}{p^2}\right\}\frac{\mathrm{d}p}{\mathrm{d}x},$$

i.e.

$$\left\{\frac{x}{x+1}-\frac{(x+1)\,\mathrm{e}^x}{p^2}\right\}\left\{\frac{\mathrm{d}p}{\mathrm{d}x}-\frac{x+2}{x+1}p\right\}=0.$$

The second factor gives a differential equation whose general integral is

$$p=C(x+1)\,\mathrm{e}^x;$$

substituting for p in the original we have its general integral

$$y=Cx\,\mathrm{e}^x+1/C.$$

The first factor gives

$$p^2=(x+1)^2\,\mathrm{e}^x/x.$$

On the one hand, integration of this gives $y=2\sqrt{(x\,\mathrm{e}^x)}+\mathit{const.}$; on the other, substitution in the original equation gives

$$y^2=4x\,\mathrm{e}^x.$$

This is therefore a singular solution; it may be verified to represent the envelope of the integral curves.

§22 Equations with one variable missing

If an equation of either of the forms

$$F(x, p)=0 \qquad \text{or} \qquad G(y, p)=0$$

is soluble for p, thus

$$p=f(x) \qquad \text{or} \qquad p=g(y)$$

integration by quadratures is immediate.

On the other hand, we may encounter equations in which solution for p is impracticable, but solution for x, or y, is possible. If for instance, we have

$$x=\phi(p) \qquad \text{or} \qquad y=\phi(p),$$

the second variable (y or x) is expressible in terms of p by quadratures. Thus if $x=\phi(p)$ we have

$$y-c=\int p\,\mathrm{d}x=\int p\phi'(p)\,\mathrm{d}p$$

or alternatively

$$y-c=\int p\,\mathrm{d}x=px-\int x\,\mathrm{d}p=px-\int \phi(p)\,\mathrm{d}p$$

and if $y=\psi(p)$ we have

$$x-c=\int \mathrm{d}y/p=\int \psi'(p)\,\mathrm{d}p/p$$

or

$$x-c=\int \mathrm{d}y/p=y/p+\int y\,\mathrm{d}p/p^2=y/p+\int \psi(p)\,\mathrm{d}p/p^2.$$

Another method is to replace the given differential equation by its equivalent in terms of a parameter t. For instance suppose that $G(y, p)=0$ is equivalent to the pair of equations

$$y=\phi(t), \qquad p=\psi(t).$$

Then

$$x-c=\int \frac{\mathrm{d}y}{p}=\int \frac{\phi'(t)\,\mathrm{d}t}{\psi(t)}$$

which, taken with $y=\phi(t)$, gives a parametric representation of the general integral.

Example 22.1

$$x^2 = p^2(a^2 - x^2).$$

Solving for x,

$$x = ap/\sqrt{(1+p^2)}; \qquad \text{i.e. } \frac{dx}{dp} = a/(1+p^2)^{3/2}.$$

Hence

$$y - c = \int p\, dx = a\int \frac{p\, dp}{(1+p^2)^{3/2}} = -\frac{a}{\sqrt{(1+p^2)}}.$$

In this case p may be eliminated, giving the general integral

$$x^2 + (y-c)^2 = a^2.$$

Example 22.2

$$\begin{aligned} y &= p^2/(p+1). \\ x - c &= \int dy/p = y/p + \int y\, dp/p^2 \\ &= y/p + \int dp/(p+1) = y/p + \log|p+1|. \end{aligned}$$

Example 22.3

$$y^2 + p^2 = a^2.$$

This is equivalent to

$$y = a \sin t, \qquad p = a \cos t,$$

Hence, since

$$\frac{dx}{dt} = \frac{1}{p}\frac{dy}{dt}$$

we have

$$\frac{dx}{dt} = 1$$

and hence

$$x - c = t.$$

The general integral is therefore

$$y = a \sin(x - c).$$

§23 Homogeneous equations

Just as a homogeneous equation of the first degree can be reduced to integrable form by the substitution $y=vx$, so also can an equation of the form $F(x, y, p)=0$, where F is homogeneous with respect to x and y. For if the degree of homogeneity is m, $F(x, vx, p)=x^m F(1, v, p)=x^m G(v, p)$ say, and the equation becomes

$$G(v, p)=0.$$

If it can be solved for p, say $p=\phi(v)$, we have

$$v+xv'(x)=\phi(v),$$

and proceed as in §4. On the other hand, if it is soluble for v, say $v=\psi(p)$, we write

$$y=vx=x\psi(p),$$

whence, differentiation with respect to x,

$$p=\psi(p)+x\psi'(p)p'(x)$$

i.e.

$$\log x=\int\frac{\psi'(p)\,\mathrm{d}p}{p-\psi(p)}$$

and, integrating, we obtain x and thence y in terms of the parameter p.

Example 23.1

$$x^2-k^2y^2+2xyp+(1-k^2)y^2p^2=0.$$

This equation is homogeneous in x and y; writing $y=vx$, cancelling the factor x^2, and solving for p, we have

$$x\frac{\mathrm{d}v}{\mathrm{d}x}+v=p=\frac{1+k\sqrt{\{1-(k^2-1)v^2\}}}{(k^2-1)v}.$$

The variables v and x are separable; writing $1-(k^2-1)v^2=z^2$ we find that the equation reduces to

$$\frac{\mathrm{d}z}{\mathrm{d}x}+\frac{z+k}{x}=0$$

whose solution is

$$x(z+k)=c,$$

i.e.

$$x^2z^2=(c-kx)^2$$

which leads to the solution

$$x^2-(k^2-1)y^2=(c-kx)^2.$$

This general integral may be written

$$x^2(1-k^2)+y^2(1-k^2)+2kcx-c^2=0$$

representing a family of circles. As c is quite arbitrary nothing is lost by regarding k as positive throughout.

§24 Geometrical interpretation of a differential equation

It will be assumed that $F(x, y, y')$ is a polynomial of degree m in y', and that each coefficient in the polynomial is a one-valued function of x and y.

Let us replace y' by z and interpret x, y, z as coordinates in space referred to rectangular axes, with the z-axis vertical. The xy-plane will then be spoken of as the horizontal plane. Consider any point A of coordinates (x_0, y_0) in this plane, such that the vertical line through A intersects the surface $F(x, y, z)=0$ in at least one point P; let the height of P above the horizontal plane be z_0. Thus the point A has a definite direction $y'=z_0$ associated with it; if A begins to move forward in this direction, P will begin to move along the surface, but the altering value of z will involve a change in the direction of motion of A.

If, then, we suppose that A moves along the horizontal plane in such a way that the direction of its motion is measured by $AP(y'=z)$, the path traced out by A will be an integral curve of the equation, for it will be a continuous curve such that at every point the relation $F(x, y, y')=0$ is satisfied. This integral curve is the horizontal projection of a certain curve on the surface $F(x, y, z)=0$, such that the condition $y'=z$ is satisfied throughout its length.

In passing, it may be noted that in the case of the Clairaut equation $y=px+f(p)$, the non-singular integral curves are the projections of the intersections of the surface $y=zx+f(z)$ by the family of parallel planes $z=c$.

To return to the point $A(x_0, y_0)$; a vertical line through A will in general intersect the surface not in one, but in several (though at most m)

distinct points. With each of these points is associated a value of z, its height above the horizontal plane, and thus a definite direction y' at A. Consequently, when the points P on the vertical through A are simple intersections with the surface, i.e. when the roots of the equation $F(x_0, y_0, z)=0$ are all distinct, integral curves, all with different tangents, will pass through A, equal in number to the real roots of this equation.

Thus through any point (x_0, y_0) of the horizontal plane there will pass at most m integral curves with distinct tangents.

Now consider a point $A(x_0, y_0)$ in the horizontal plane, such that the equation $F(x_0, y_0, z)=0$ has a repeated root. Then we should naturally expect two or more of the integral curves that pass through A to have the same tangent at A. We shall investigate this case more closely, with a view to ascertaining whether or not this supposition agrees with fact.

The analytic condition for the equation

$$F(x_0, y_0, z)=0$$

to have a double or multiple root is that

$$\frac{\partial}{\partial z} F(x_0, y_0, z)=0$$

simultaneously with it. The geometrical condition is that the vertical line through A shall meet the surface in two or more coincident points, i.e. shall be a tangent line to the surface.

Let A move so that the vertical line continues to touch the surface; that is to say, in such a way that the equations

$$F(x, y, z)=0, \qquad \frac{\partial}{\partial z} F(x, y, z)=0 \tag{24.1}$$

are simultaneously satisfied. Then the path of A is the trace on the horizontal plane of a vertical cylinder enveloping the surface. The most obvious cylinder is that which touches the surface along what is known in descriptive geometry as the visible outline in horizontal projection; its trace forms a natural boundary to the family of integral curves, but, as will be seen, is not necessarily an integral curve itself. There may be other vertical cylinders touching the surface along a curve; their traces generally cut across the family of integral curves. All possible traces are included in the eliminant of (24.1), which represents a curve or curves, known as the *discriminant locus*.

For any infinitesimal displacement (dx, dy, dz) on the surface, we have

$$F_x\,dx + F_y\,dy + F_z\,dz = 0,$$

(where a suffix denotes partial differentiation with respect to the variable in question) but since at any point on the curve of contact of an enveloping cylinder, $F_z=0$ (24.1), it follows that for any such displacement originating on the curve of contact

$$F_x\,\mathrm{d}x+F_y\,\mathrm{d}y=0. \tag{24.2}$$

For a displacement along an integral curve, we have

$$-z\,\mathrm{d}x+\mathrm{d}y=0. \tag{24.3}$$

If the horizontal projection of the displacement on the surface, at a point (x, y, z) on a curve of contact, is along an integral curve, (24.2) must be consistent with (24.3). Consequently, either $\mathrm{d}x=\mathrm{d}y=0$, which implies that the displacement on the surface is vertical, or

$$\begin{vmatrix} F_x & F_y \\ -z & 1 \end{vmatrix}=0,$$

that is to say,

$$\frac{\partial F}{\partial x}+z\frac{\partial F}{\partial y}=0 \tag{24.4}$$

in conjunction with (24.1).

§25 Cusp on the integral curve

Let us suppose that the equation of the integral curve of $F(x, y, y')=0$ that passes through (x_0, y_0) can be expressed in terms of a parameter t, thus

$$x=x(t), \qquad y=y(t).$$

This integral curve is the projection of a curve

$$x=x(t), \qquad y=y(t), \qquad z=z(t)$$

on the surface, with the condition $\mathrm{d}y=z\,\mathrm{d}x$. There will be no loss in generality if we assume that the point (x_0, y_0, z_0) on the curve of contact corresponds to the value $t=0$.

We shall consider the case of a vertical displacement on the surface from the point (x_0, y_0, z_0), i.e. $\mathrm{d}x=\mathrm{d}y=0$ when $t=0$; hence

$$x-x_0=\tfrac{1}{2}x''(0)t^2+\tfrac{1}{6}x'''(0)t^3+\cdots$$

$$y-y_0=\tfrac{1}{2}y''(0)t^2+\tfrac{1}{6}y'''(0)t^3+\cdots$$

$$z-z_0=z'(0)t+\cdots$$

The approximation to the integral curve in the neighbourhood of (x_0, y_0) is therefore

$$x''(0)(y-y_0)-y''(0)(x-x_0)=\tfrac{1}{6}\{x''(0)y'''(0)-y''(0)x'''(0)\}t^3+\cdots$$
$$=K(x-x_0)^{3/2}+\cdots$$

where K is a constant. Thus the integral curve has a *cusp* at the point (x_0, y_0) on the discriminant locus, and the direction of the tangent at the cusp is $y''(0)/x''(0)$. But the direction of the tangent to the discriminant locus itself at (x_0, y_0) is $(-F_x/F_y)_0$. These directions are in general unrelated so that the integral curve does not touch the outline. Thus in this most general case, the discriminant locus is a cusp locus. A more particular case, that of a tac locus, will be discussed briefly in a later section (§27).

§26 Envelope of integral curves

Suppose the condition $dx=dy=0$ is not satisfied, which (24.4) implies that

$$F_x+zF_y=0$$

at all points of the curve of contact on the surface. But we also have the condition (24.2)

$$F_x\,dx+F_y\,dy=0,$$

where dy/dx is the direction of a displacement along the discriminant locus. Therefore, this value of dy/dx may be identified with z in $F(x, y, z)=0$; that is to say, the portion of the discriminant locus under consideration is an integral curve of the differential equation $F(x, y, y')=0$. Thus when the curves on the surface which project into the integral curves do not cross the curve of contact vertically, the latter curve projects into a branch of the discriminant locus which is itself an integral curve. If (x, y, z) is a point on the curve of contact, (x, y) is a point both on the discriminant curve and on one of the general family of integral curves, and these two curves have a common value of y' at the point, i.e. they have a common tangent. Thus at every point on the discriminant curve in question it is in contact with one of the general integral curves, and is therefore an envelope of the family of integral curves. In very special cases the envelope is itself a member of the general family, but in general it is not; it is then known as a singular integral curve, representing a singular solution of the equation. To sum up, reverting to the p-notation:

The equation obtained by eliminating p between

$$F(x, y, p)=0 \quad \text{and} \quad F_p(x, y, p)=0 \tag{26.1}$$

is known as the p-discriminant equation; it represents a curve (the *p-discriminant locus*) of one or more branches in the plane of the general integral curves. A branch is a singular integral curve, i.e. an envelope, if and only if the further condition

$$F_x(x, y, p)+pF_y(x, y, p)=0 \tag{26.2}$$

holds at every point. Otherwise the branch is a locus of special points, generally of cusps, on the integral curves.

Note 1. In the case of the Clairaut equation $px-y+f(p)=0$ or the generalized equation $f(p)+g(px-y)=0$, condition (26.2) is identically satisfied, so that the p-discriminant furnishes the singular solution and nothing else.

Note 2. In the case of $f(p)+xg(p)=0$ (20.1) the second of equations (26.1) gives $f'(p)+xg'(p)=0$, which implies (20.2) $g(p)-p=0$. But this is precisely the condition imposed by (26.2), showing that the singular integral curves are straight lines whose gradients are given by the real roots of $g(p)-p=0$.

§27 Equation of the second degree

Let (x_0, y_0, z_0) be a point on the curve of contact of the surface $F(x, y, z)=0$ with its vertical enveloping cylinder, so that $z=z_0$ is a double root of the equation $F(x_0, y_0, z)=0$, which implies that $F(x_0, y_0, z)$ has the factor $(z-z_0)^2$. For values of x, y sufficiently near to $x_0, y_0, F(x, y, z)$ has two factors $z-z_1, z-z_2$, where z_1, z_2 are functions of x, y which both become equal to z_0 when $x=x_0$, $y=y_0$. Thus the factor $(z-z_0)^2$ is the limit, as $x\to x_0$, $y\to y_0$, of a quadratic expression $z^2-(z_1+z_2)z+z_1z_2$, and therefore, in the neighbourhood of the p-discriminant, the differential equation $F(x, y, p)=0$ may be regarded as approximated to by a quadratic differential equation

$$p^2-(z_1+z_2)p+z_1z_2=0,$$

and thus consideration of such an equation will confirm, and possibly supplement, the discussion of the preceding sections.

We consider, therefore, the following general type of equation of the second degree

$$p^2L(x, y)-2pM(x, y)+N(x, y)=0, \qquad (27.1)$$

and we shall assume that L, M, N are functions developable as ascending power series in x and y. Solving for p, we have

$$p=\{M\pm\surd(M^2-LN)\}/L \qquad (27.2)$$

Thus the xy-plane is divided into distinct regions, namely:

(*a*) region for which $M^2<LN$ in which no integral curves exist;

(*b*) regions for which $M^2>LN$, where two distinct real values of p exist for every (x, y), i.e. two integral curves with distinct tangents pass through every point.

The two values of p are equal for every point at which

$$M^2-LN=0.$$

Now this equation represents a curve which, however, may be composed of several distinct branches. Thus it includes

(*c*) the frontier between the regions (*a*) and (*b*) characterized by the fact that passage across the frontier involves a change in the sign of M^2-LN;

(*d*) curves situated within the regions (*a*) or (*b*) such that M^2-LN vanishes without a change of sign.

Cusp locus. Let the origin 0 be moved to a point on a branch Γ of the discriminant locus which is a frontier between regions (*a*) and (*b*). Then in the neighbourhood of 0,

$$L=l_0+l_1x+l_2y+\cdots$$

$$M=m_0+m_1x+m_2y+\cdots$$

$$N=n_0+n_1x+n_2y+\cdots$$

and $m_0^2=l_0n_0$.

Then the gradient of the integral curve that goes through O is (27.2)

$$p=m_0/l_0=n_0/m_0.$$

But near the origin

$$M^2-LN=m_0^2+2m_0(m_1x+m_2y)+\cdots$$

$$=l_0n_0-l_0(n_1x+n_2y)-n_0(l_1x+l_2y)-\cdots,$$

so that the tangent to Γ at the origin has the equation

$$(2m_0m_1 - l_0n_1 - n_0l_1)x + (2m_0m_2 - l_0n_2 - n_0l_2)y = 0$$

or, say,

$$ax + by = 0.$$

Since $M^2 - LN$ changes sign we may assume that one at least of a and b is not zero. Since $-a/b$ differs in general from m_0/l_0, the slope of Γ at 0 differs from that of the integral curve, so that Γ is not itself an integral curve. To find how the integral curve through 0 behaves with respect to Γ, we require a further approximation to (27.2):

$$p = \frac{m_0 + m_1x + m_2y + \cdots \pm \sqrt{(ax + by + \cdots)}}{l_0 + l_1x + l_2y + \cdots}.$$

The first approximation is $p = m_0/l_0$; we therefore write $y = m_0x/l_0 + Y$, where Y is of higher degree than the first in x, and obtain an equation of the form

$$\frac{dY}{dx} = \frac{\alpha x + \cdots \pm \sqrt{(\beta x + \cdots)}}{l_0 + \cdots} = \pm \gamma x^{1/2} + \cdots,$$

where α, β, γ are constants. Integrating, we have

$$Y = \pm \tfrac{2}{3}\gamma x^{3/2} + \cdots$$

or

$$y = m_0x/l_0 \pm \tfrac{2}{3}\gamma x^{3/2} + \cdots,$$

so that the integral curve has a cusp on Γ. Hence any branch of the discriminant locus that separates a region of existence from a region of non-existence of integral curves is *in general* a locus of cusps on the integral curves.

Envelope. Now consider the special case when $-a/b = m_0/l_0$. When the substitution

$$y = m_0x/l_0 + Y = -ax/b + Y$$

is made, the term in x under the radical disappears, leaving

$$\frac{dY}{dx} = \frac{ax + \cdots + \sqrt{(cx^2 + bY + \cdots)}}{l_9 + \cdots}.$$

It will be found impossible to solve this equation by $Y = \lambda x^{\mu} + \cdots$ with $\mu < 2$; hence it is of the form

$$Y' = (\beta \pm \gamma)x + \cdots,$$

which gives

$$y = m_0 x/l_0 + \tfrac{1}{2}(\beta \pm \gamma)x^2 + \cdots,$$

showing that two distinct integral curves touch one another at 0. When this happens continuously along a branch of the discriminant curve, since the slope of the latter is everywhere the slope of an integral curve, it is itself an integral curve. So in this case one of the curves having contact is a branch Γ of the discriminant curve, the other is one of the family of integral curves, which are therefore enveloped by Γ.

An application of (26.2) at 0 gives

$$(p^2 l_1 - 2pm_1 + n_1) + p(p^2 l_2 - 2pm_2 + n_2) = 0.$$

If each bracket is divided by p and if p is replaced by n_0/m_0 and $1/p$ by l_0/m_0 this reduces to $a + pb = 0$, and therefore if (26.2) holds at any point of Γ, Γ there touches an integral curve; if it holds at all points of Γ, then Γ is an envelope of the integral curves.

Tac locus. Now suppose that 0 is moved to a branch Γ of the discriminant curve for which $M^2 - LN$ vanishes without changing sign; the sign on either side of Γ will be assumed to be positive. Then in the neighbourhood of 0

$$M^2 - LN = (ax + by)^2 + \cdots.$$

The form of the integral curve through 0 is given approximately by

$$p = \frac{m_0 + m_1 x + m_2 y + \cdots \pm \sqrt{\{(ax+by)^2 + \cdots\}}}{l_0 + l_1 x + l_2 y + \cdots},$$

from which we find that

$$y = m_0 x/l_0 + (\alpha + \beta)x^2 + \cdots.$$

Thus two distinct integral curves touch one another on the curve Γ. Since the same is true at all its points, Γ is a *tac locus.* In general $-a/b$ is not equal to m_0/l_0 and therefore the tac locus is not an integral curve.

Note 1. The above assumes $M^2 - LN$ to be positive, and the integral curves real on either side of the tac locus. In the contrary case when $M^2 - LN$ is negative, there is a real tac locus of imaginary integral curves.

Note 2. Since the approximation to the tac locus, at any point on it, is a squared term, the tac locus occurs as a squared factor in the p-discriminant.

Example 27.1

$$xp^2+(y-3x)p+my=0.$$

This equation may be solved for y and integrated by the method of §20. The p-discriminant equation is $(y-3x)^2-4mxy=0$; the corresponding curve consists of two straight lines through the origin which are real except when $0>m>-3$, and are coincident when $m=0$ or -3. The condition for a singular integral (26.2) gives

$$p^2-3p+p(p+m)=0.$$

The first root $p=0$ would lead to $y=0$, which is no part of the p-discriminant; the second root $2p=3-m$ gives $2y=(3-m)x$. On substituting this in the p-discriminant equation, we find $m=1$ (twice) to be the only admissible case, giving $p=1$, and note that $y=x$ is a solution of the differential equation.

Integrating the differential equation for $m=1$ in terms of the parameter p, we have

$$x^2=c(p+1)^2/p^3, \qquad y^2=c(p-3)^2/p$$

or, eliminating p between this expression for y^2 and the differential equation, we obtain the general integral

$$(xy^2+cy+3cx)(y^3+15cy-27cx)+c^2(y-9x)^2=0.$$

We find that $y=x$ has, in fact, contact with all integral curves for which $c>0$, touching each at two distinct points $x=y=\pm\sqrt{c}$. This contact at two points accounts for the double value of m; it is as if the family of integral curves had two coincident envelopes. Now consider the other branch, $y=9x$, of the p-discriminant locus; squaring and substituting the above parametric values of x, y we find that the values of p at the points at which $y=9x$ meets any integral curves are given by

$$(p^2-12p-9)(p^2+6p+9)=0.$$

The two single values from the first bracket are of no interest, they merely correspond to simple intersections. The double value $p=-3$, however, shows that every integral curve encounters $y=9x$ in points $x^2=-4c/27$ at which it has a double tangent. To investigate the behaviour of the integral curve near such a point we put $p=-3+t$, where t is small, in the parametric expressions for x and y. If $k=\pm\sqrt{(-27/4c)}$ we find

$$x=c^{1/2}(-2+t)(-3+t)^{-3/2}$$

or

$$kx = 1 - \tfrac{1}{24}t^2 - \tfrac{5}{216}t^3 + \cdots$$

$$(3p - p^2)/(p+1) = 9(1 - \tfrac{1}{2}t + \tfrac{1}{18}t^2)(1 - \tfrac{1}{2}t)^{-1}$$

$$= 9(1 + \tfrac{1}{18}t^2 + \tfrac{1}{216}t^3 + \cdots)$$

$$ky = kx(3p - p^2)/(p+1) = 9 + \tfrac{1}{8}t^2 + \tfrac{1}{24}t^3 + \cdots.$$

Hence

$$3(kx-1) + (ky-9) = -\tfrac{1}{36}t^3 + \cdots$$

and

$$\{3(kx-1) + (ky-9)\}^2 = \tfrac{32}{3}(kx-1)^3 + \cdots.$$

Thus every integral curve for which c is negative has a cusp on the line $y = 9x$, the double tangent at the cusp being parallel to $3x + y = 0$. This is an instance where a branch of the p-discriminant locus is a cusp locus.

To illustrate the circumstances that arise when the p-discriminant locus has a double line, take the case $m = -3$. The general integral is now

$$(xy - c)(y - 3x + c) = 0$$

and thus consists of a family of rectangular hyperbolas and a family of parallel straight lines taken together. The p-discriminant locus is the double line $(y + 3x)^2 = 0$. Now every one of the parallel straight lines touches one hyperpola of the family; in fact $y = 3x - 6\gamma$ touches $xy + 3\gamma^2 = 0$ at the point $x = \gamma$, $y = -3\gamma$ which is a point on the discriminant line. This is a tac point, or point of contact of two distinct integral curves, and the double line of the p-discriminant is a tac locus.

4

Equations of the second and higher orders

§28 Reduction of the order of an equation

When we turn to equations of higher order than the first, we discover that there exist certain well-defined types which admit of transformations whereby the order may be lowered. In particular, there are equations that may be reduced, by a transformation of the dependent variable, to allied equations of the first order; if the latter can be integrated, a reversal of the transformation (which usually amounts to one or more quadratures) leads to the general integral of the former.

There is one small point that may be noted in passing, namely that whereas in an equation of the first order, e.g. $P+Qy'=0$, there is no natural discrimination between the variables, in equations where second and higher derivatives of one variable (y) with respect to the other (x) occur, the distinction is evident. In equations that arise from problems in physics and mechanics, in particular, this distinction arises from a difference in the character of the variable; as when one represents a length and the other represents time.

We shall deal mainly with equations of the second order and shall consider, in particular, those classes in which a change of dependent variable enables a reduction to the first order to be effected. No general discussion of higher orders will be attempted, but when the scope of any process extends beyond the second order, the fact will be mentioned.

The most general differential equation of the second order in which x is independent, and y dependent variable, may be written

$$F(x, y, y', y'')=0. \tag{28.1}$$

If it may be derived from an equation

$$f(x, y, c_1, c_2)=0 \tag{28.2}$$

in which c_1, c_2 are arbitrary constants, by differentiating twice with respect to x and eliminating these two constants, (28.2) is known as its primitive. The elimination of c_1 and c_2 may be performed in either order, but can only lead to the one equation (28.1). For if two such equations were found, y'' could be eliminated between them, leaving an equation of the first order having the primitive (28.2), which cannot be the case unless c_1 and c_2 are specially related. Integration consists of recovering (28.2) or any equivalent expression containing two arbitrary constants, which is a general integral.

§29 Equations that do not involve y

The simplest case where reduction of order is possible occurs when the dependent variable y itself is absent from the equation, which may be written

$$F(x, y', y'')=0. \tag{29.1}$$

We replace y' by p and regard p as a new dependent variable, temporarily replacing y, and thus have

$$F(x, p, p')=0, \tag{29.2}$$

an equation of the first order in p.

Let us suppose for the moment that this equation can be integrated explicitly, thus

$$p=f(x, c_1), \tag{29.3}$$

introducing one arbitrary constant c_1. We then obtain y by the quadrature

$$y=\int p\,\mathrm{d}x=\int f(x, c_1)\,\mathrm{d}x+c_2, \tag{29.4}$$

introducing the second arbitrary constant c_2 and thus arriving at the general integral.

Suppose, on the other hand, the integration of (29.2) gives x more naturally in terms of p, so that we have

$$x=g(p, c_1). \tag{29.5}$$

We proceed to take the derivative of both sides of this equation to obtain

$$\frac{\mathrm{d}x}{\mathrm{d}p}=g'(p, c_1)$$

and thus we have

$$y=\int p\,\mathrm{d}x=\int pg'(p, c_1)\,\mathrm{d}p+c_2. \tag{29.6}$$

or, alternatively, we may write

$$y=\int p\,dx=px-\int x\,dp=pg(p,c_1)-\int g(p,c_1)\,dp+c_2.$$

Thus x and y are expressed in terms of the parameter p.

When x is absent as well as y, and the equation can be written as

$$y''=f(y')$$

the process is simplified, for we have

$$p'=f(p) \quad \text{or} \quad \frac{dx}{dp}=\frac{1}{f(p)}.$$

Also

$$\frac{dy}{dp}=p\frac{dx}{dp}=\frac{p}{f(p)}$$

and the general integral is given parametrically by

$$x=\int dp/f(p)+c_1, \qquad y=\int p\,dp/f(p)+c_2. \tag{29.7}$$

In the case of the equation of order n

$$y^{(n)}=f(y^{(n-1)})$$

we may write $z=y^{(n-1)}$ and thus obtain as in (29.7)

$$x=\int dz/f(z)+c_1, \qquad y^{(n-2)}=\int z\,dz/f(z)+c_2$$

and then in turn

$$y^{(n-3)}=\int y^{(n-2)}\,dx=\int y^{(n-2)}\,dz/f(z)$$

$$=\int\frac{dz}{f(z)}\int z\frac{dz}{f(z)}+c_1x+c_2$$

$$y^{(n-4)}=\int\frac{dz}{f(z)}\int\frac{dz}{f(z)}\int z\frac{dz}{f(z)}+\tfrac{1}{2}c_1x^2+c_2x+c_3$$

and so on until we arrive at y, which will involve an arbitrary polynomial of degree $n-2$ (involving $n-1$ arbitrary constants).

An equation of the type $F(x, y^{(n-1)}, y^{(n)})=0$ may also be reduced to the first order by writing $z=y^{(n-1)}$.

Example 29.1

$$x^2y''=y'^2-2xy'+2x^2.$$

Writing $y'=p$, we have

$$x^2p'=p^2-2xp+2x^2,$$

an equation homogeneous in x and p; if $p=vx$ it reduces to

$$xv'=v^2-3v+2$$

whose integral is

$c_1x=(v-2)/(v-1)$, whence $v=(c_1x-2)/(c_1x-1)$. Thus

$$\frac{\mathrm{d}y}{\mathrm{d}x}=xv=\frac{x(c_1x-2)}{c_1x-1}=x-\frac{1}{c_1}\left\{1+\frac{1}{c_1x-1}\right\}$$

and finally

$$y=\tfrac{1}{2}x^2-\frac{x}{c_1}-\frac{1}{c_1^2}\log|c_1x-1|+c_2.$$

Example 29.2

$$2x^2y'y''-xy''+y'=0.$$

Writing $y'=p$ and dividing by x^2 we have

$$2pp'-\frac{p'}{x}+\frac{p}{x^2}=0,$$

which is exact and has the integral

$$p^2-\frac{p}{x}+c_1=0.$$

We thus obtain

$$x=\frac{p}{p^2+c_1},$$

whence

$$y=\int p\,\mathrm{d}x=px-\int x\,\mathrm{d}p$$

$$=\frac{p^2}{p^2+c_1}-\int\frac{p\,\mathrm{d}p}{p^2+c_1}=\frac{p^2}{p^2+c_1}-\tfrac{1}{2}\log|p^2+c_1|+c_2.$$

It is here possible to eliminate p.

Example 29.3

$$xy''' - 2y'' = x^3.$$

The substitution $y'' = z$ gives the first-order linear equation

$$xz' - 2z = x^3,$$

whose solution leads to the equation

$$y'' = z = x^3 + c_1 x^2$$

and in turn

$$y' = \tfrac{1}{4}x^4 + \tfrac{1}{3}c_1 x^3 + c_2$$

$$y = \tfrac{1}{20}x^5 + \tfrac{1}{12}c_1 x^4 + c_2 x + c_3.$$

§30 Equations that do not involve x.

When the independent variable is lacking, the equation takes the form

$$F(y, y', y'') = 0, \tag{30.1}$$

and we again write p for y'. But as y, p are now the variables involved, y'' requires to be transformed as follows

$$y'' = \frac{dp}{dx} = \frac{dp}{dy}\frac{dy}{dx} = p\frac{dp}{dy}.$$

The equation then becomes

$$F(y, p, p\, dp/dy) = 0; \tag{30.2}$$

it is of the first order with y as dependent variable. If an integral of the form

$$p = f(y, c_1)$$

is obtainable, we separate the variables, and obtain the general integral

$$x = \int \frac{dy}{f(y, c_1)} + c_2.$$

On the other hand, when the first integral appears in the form

$$y = g(p, c_1),$$

we have

$$x=\int\frac{\mathrm{d}y}{p}=\int\frac{g'(p,c_1)\,\mathrm{d}p}{p}+c_2$$

or

$$x=\int\frac{\mathrm{d}y}{p}=\frac{y}{p}+\int\frac{y\,\mathrm{d}p}{p^2}=\frac{y}{p}+\int\frac{g(p,c_1)\,\mathrm{d}p}{p^2}+c_2.$$

If y' is absent from an equation of the above type, and it can be expressed in the form

$$y''=f(y),$$

we write $y''=p\,\mathrm{d}p/\mathrm{d}y$ and obtain

$$p\frac{\mathrm{d}p}{\mathrm{d}y}=f(y).$$

Integrating, we obtain the equation

$$p^2=2\int f(y)\,\mathrm{d}y+c_1$$

whence

$$x=\int\frac{\mathrm{d}y}{p}=\int\frac{\mathrm{d}y}{\sqrt{\{2\int f(y)\,\mathrm{d}y+c_1\}}}+c_2.$$

An equation that involves three consecutive derivatives, and nothing else, say $F(y^{(n-2)}, y^{(n-1)}, y_n)=0$, may similarly be reduced by taking $y^{(n-2)}=u$, $y^{(n-1)}=z$, $y^{(n)}=z\,\mathrm{d}z/\mathrm{d}u$.

Example 30.1

$$y(y-1)y''+y'^2=0.$$

The transformed equation is

$$y(y-1)pp'(y)+p^2=0.$$

The variables p and y are separable and the first integral is

$$c_1 p=y/(y-1).$$

Hence

$$x=\int\frac{\mathrm{d}y}{p}=c_1\int\frac{y-1}{y}\,\mathrm{d}y=c_1 y-c_1\log|y|+c_2.$$

Example 30.2

$$y'' + m^2 y = 0.$$

This becomes

$$pp'(y) + m^2 y = 0.$$

Separating the variables and integrating, we have

$$p^2 + m^2 y^2 = m^2 c_1^2,$$

whence

$$p = m\sqrt{(c_1^2 - y^2)}$$

or

$$\int \frac{\mathrm{d}y}{\sqrt{(c_1^2 - y^2)}} = \pm \int m\,\mathrm{d}x$$

and integrating

$$\arcsin(y/c_1) = \pm mx + c_2,$$

i.e.

$$y = c_1 \sin(c_2 \pm mx)$$

which may also be written as $y = A\cos mx + B\sin mx$.

Example 30.3

$$y^{(n-2)} y^{(n)} = \{y^{(n-1)}\}^2.$$

The transformation mentioned above gives

$$uzz'(u) = z^2$$

whence we deduce that

$$z = c_1 u \quad \text{or} \quad u' = c_1 u.$$

Thus we have

$$y^{(n-2)} = u = c_2\,\mathrm{e}^{c_1 x}$$

and integrating $n-2$ times in succession we obtain the general integral in the form

$$y = K\,\mathrm{e}^{cx} + \text{an arbitrary polynomial of degree } n-3.$$

§31 First homogeneous type

The expression

$$F(x, y, y', y'')$$

is said to be homogeneous and of degree m in y and its derivatives if, when λ is any constant,

$$F(x, \lambda y, \lambda y', \lambda y'') = \lambda^m F(x, y, y', y''). \tag{31.1}$$

When this is the case, the equation $F(x, y, y', y'') = 0$ may be reduced, by division throughout by y^m, to one still homogeneous in y, y', y'', but of degree zero. Its form is then

$$f(x, y'/y, y''/y) = 0. \tag{31.2}$$

The order may now be lowered by taking $y'/y = u$ as a new dependent variable, i.e. by the transformation

$$y = \exp\{\int u\,dx\}$$

which implies $y' = uy$, $y'' = y(u' + u^2)$. The equation then becomes

$$f(x, u, u' + u^2) = 0. \tag{31.3}$$

It is now of the first order, but here as in most cases the lowering of order is paid for by an increase in complexity. Thus, when the method is applied to the linear equation

$$y'' + p(x)y' + q(x)y = 0,$$

which is homogeneous and of the first degree in y, y', y'', it becomes

$$u' + u^2 + p(x)u + q(x) = 0,$$

a Riccati equation (§12) which is actually less manageable than the equivalent linear equation.

When an equation of order n is expressible in the form $f(x, y'/y, \ldots, y^{(n)}/y) = 0$, it may be transformed by the above process into an equation of order $n - 1$.

Example 31.1

$$xyy'' - xy'^2 + yy' = 0 \qquad \text{(second degree in } y, y', y''\text{)}.$$

This may be written

$$x\frac{y''}{y} - x\left(\frac{y'}{y}\right)^2 + \frac{y'}{y} = 0.$$

The transformed equation is

$$x(u'+u^2)-xu^2+u=0 \qquad \text{or} \qquad xu'+u=0.$$

Thus we obtain the first integral

$$xu=a$$

or

$$xy'=ay$$

and finally arrive at the general integral

$$y=cx^a,$$

where a and c are constants.

§32 Second homogeneous type

In this case the equation is of the form

$$f(y, xy', x^2y'')=0. \tag{32.1}$$

We reduce it to a simpler form by the substitution

$$x=\mathrm{e}^t$$

so that

$$y'(x)=\dot{y}/\dot{x}=\mathrm{e}^{-t}\,\dot{y}$$

where $\dot{y}=\mathrm{d}y/\mathrm{d}t$ etc. From this we deduce that

$$xy'(x)=\dot{y}(t)$$

and hence that

$$x\frac{\mathrm{d}}{\mathrm{d}x}\{xy'(x)\}=\ddot{y}(t)$$

i.e. that

$$x^2y''(x)=\ddot{y}(t)-\dot{y}(t).$$

Equation (32.1) is then reduced to

$$f(y, \dot{y}, \ddot{y}-\dot{y})=0 \tag{32.2}$$

The equation can be further reduced to a first-order equation by writing

$$\dot{y}=v$$

and, as in §30,

$$\ddot{y}=vv'(y).$$

Equation (32.2) – and hence equation (32.1) – can, in this way, be reduced to the first-order equation

$$f(y, v, vv'(y)-v)=0. \tag{32.3}$$

An equation of order n which is expressible in the form $f(y, xy', \ldots, x^n y^{(n)})=0$ may be reduced in the same way to an equation of order $n-1$.

Example 32.1

$$xy''(x^2y''+2xy'+2y)+2yy'=0.$$

Changing the variables as above, we have

$$\left(v\frac{dv}{dy}-v\right)\left(v\frac{dv}{dy}+v+2y\right)+2yv=0$$

which reduces to

$$v\left(\frac{dv}{dy}\right)^2+2y\frac{dv}{dy}-v=0.$$

If v^2 is taken as dependent variable, this equation becomes of Clairaut type; its integral is

$$v^2=4a(y+a)$$

where a is a constant.

Writing $v=x\,dy/dx$ and separating the variables, we have

$$\pm\int\frac{dy}{2\sqrt{(ay+a^2)}}=\log x+c$$

whence we derive the solution

$$\pm\sqrt{(ay+a^2)}=\alpha\log(cx)$$

Rationalizing, we see that the general solution may be written as

$$y=a\{(\log cx)^2-1\}.$$

§33 Third homogeneous type

This refers to an equation of the type

$$f(y/x, y', xy'')=0. \tag{33.1}$$

The transformation $y=xu$ changes it to

$$f(u, xu'+u, x^2u''+2xu')=0.$$

i.e. to an equation of the second type, which may be dealt with by the procedure outlined in §32.

Another method of reducing (33.1) to a first-order equation is based on the double change of variable

$$y=xu, \qquad v=xu'(x) \tag{33.2}$$

so that

$$y'=u+v$$

and

$$xy''=x\left(1+\frac{\mathrm{d}v}{\mathrm{d}u}\right)u'(x)=v+v\frac{\mathrm{d}v}{\mathrm{d}u}\cdot$$

Equation (33.1) is then reduced to the first-order equation

$$f(u, u+v, v+vv'(u))=0. \tag{33.3}$$

The order of any equation of the type

$$f(y/x, y', \ldots, x^{n-1}y^{(n)})=0$$

may be reduced by one in this way.

Example 33.1

$$x^3y''+m(xy'-y)^2=0.$$

Writing this equation as $xy''+m(y'-y/x)^2=0$ and making the above transformation, we have

$$v+v\frac{\mathrm{d}v}{\mathrm{d}x}+mv^2=0 \qquad \text{or} \qquad \frac{\mathrm{d}v}{\mathrm{d}u}+mv+1=0.$$

This linear equation has the integrating factor e^{mu}; we thus obtain

$$\mathrm{e}^{mu}(1+mv)=c_1$$

i.e.

$$e^{mu}\left(1+mx\frac{du}{dx}\right)=c_1 \qquad \text{or} \qquad \frac{d}{dx}(x\,e^{mu})=c_1.$$

Integrating again, we arrive at the general integral

$$x\,e^{mu}=c_1x+c_2$$

or

$$e^{my/x}=c_1+c_2/x.$$

§34 A special case of homogeneity

An equation homogeneous in y, xy' and x^2y'' may be written

$$F(xy'/y, x^2y''/y)=0 \qquad \text{or} \qquad x^2y''=yf(xy'/y). \tag{34.1}$$

It is both of the first and the second type, and therefore also of the third, and may be reduced by taking a new dependent variable u where $u=xy'/y$. Thus we have

$$xy'=uy, \qquad xy''=u'y+(u-1)y',$$

i.e.

$$x^2y''=u'xy+u(u-1)y,$$

and so the equation, in its second form, becomes

$$xu'+u(u-1)=f(u).$$

The variables are separable, giving

$$\int\frac{dx}{x}=\int\frac{du}{f(u)-u(u-1)}$$

and hence

$$\int\frac{dy}{y}=\int\frac{u\,du}{f(u)-u(u-1)}.$$

Integrating, we obtain a parametric representation of x and y in terms of u, involving two constants of integration. Alternatively we may integrate the first equation and obtain $x=\phi(u,c)=\phi(xy'/y,c)$. Solving for xy'/y we obtain an equation whose variables are separable.

Example 34.1 The Euler linear equation of the second order,

$$x^2y'' - (a+b-1)xy' + aby = 0,$$

where a, b are constants, is of this type. The above transformation leads to

$$xu' + (u-a)(u-b) = 0,$$

whence, separating variables and integrating,

$$\frac{u-a}{u-b} = c_1 x^{a-b}$$

or

$$\frac{c_1 a x^a - b x^b}{c_1 x^a - x^b} = u = \frac{xy'}{y},$$

i.e.

$$\frac{\mathrm{d}(c_1 x^a - x^b)}{c_1 x^a - x^b} = \log y - \log c_2$$

whence $y = c_2(c_1 x^a - x^b)$, that is to say the general integral may be written $y = Ax^a + Bx^b$, where A and B are arbitrary constants.

Example 34.2

$$xyy'' = y'(2xy' + ay).$$

Taking this equation in the form

$$x^2y'' = y\frac{xy'}{y}\left(2\frac{xy'}{y} + a\right)$$

and applying the same transformation, we obtain

$$xu' = u(u+a+1)$$

whence

$$\log x = \int \frac{\mathrm{d}u}{u(u+a+1)}; \qquad \log y = \int \frac{\mathrm{d}u}{u+a+1};$$

$$x^{a+1} = c_1 \frac{u}{u+a+1}; \qquad y = c_2(u+a+1).$$

By eliminating u, we see that the general integral may be expressed in the form

$$y=(A+Bx^{a+1})^{-1}.$$

§35 First integral

The term first integral has been used to denote (§30) the differential equation of the first order, involving one arbitrary constant, that results from transforming a differential equation of the second order into one of the first, and integrating the latter. When the first integral is obtainable by immediate integration of the differential equation, the latter is said to be *exact*. Thus

$$2y'y''+xy'+y=0$$

is exact and its first integral is

$$y'^2+xy=c_1.$$

In the same way, a first integral of an equation of order n

$$F(x, y, y', y'', \ldots, y^{(n)})=0 \tag{35.1}$$

will be an equation of order $n-1$

$$F_1(x, y, y', y'', \ldots, y^{(n-1)}, c_1)=0, \tag{35.2}$$

where x_1 is arbitrary. Moreover, when $\mathrm{d}F_1/\mathrm{d}x=F$, then $F=0$ is exact.

As in the case of equations of the first order, when an equation of any order is not exact as it stands, it may be rendered exactly by the introduction of an integrating factor. For example, the equation

$$y''+P(x, y)y'+Q(x, y)y'^2=0$$

admits of the integrating factor $1/y'$ whenever the equation $P+Qy'=0$ is exact, and the first integral then is

$$\log|y'|+\int(P\,\mathrm{d}x+Q\,\mathrm{d}y)=c_1.$$

The equation is also integrable (i) when P and Q are functions of x alone, for it is then equivalent to a Bernoulli equation in p, (ii) when P and Q are functions of y alone, for then it may be reduced to a linear equation in the variables y and p.

A differential equation of the second order has two distinct first integrals. For let the primitive be $f(x, y, A, B)=0$, where A, B are distinct arbitrary constants (distinct in the sense of not being replaceable by a single constant c, as would be the case, for example, if A, B occurred only in the combination $A+B$). By differentiating the primitive we obtain an equation $\phi(x, y, y', A, B)=0$, and then by eliminating B and A in turn between $f=0$, $\phi=0$ we arrive at

$$F_1(x, y, y', A)=0, \qquad F_2(x, y, y', B)=0.$$

which are formally distinct because A and B are distinct in the primitive. By differentiating again and eliminating A, B between the two resulting equations and the primitive we reach the differential equation

$$F(x, y, y', y'')=0.$$

Integrating the differential equation once results in the recovery of either $F_1=0$ or $F_2=0$ or an equivalent. Thus $F_1=0$, $F_2=0$ are first integrals, and we see that two distinct ones exist. If, by adopting two separate methods of integration we recover both $F_1=0$ and $F_2=0$, we obtain the primitive by eliminating y' between them. There cannot be more than two distinct first integrals, for if $F_3=0$ were a third, a distinct primitive would be obtained by eliminating y' between $F_1=0, F_3=0$. But the existence theorems prove, what we shall here regard as a postulate, that there cannot be more than one distinct primitive dependent upon a set of arbitrary constants equal in number of the order of the equation, i.e. there cannot be more than one distinct general integral.

Therefore an equation of the second order has just two distinct first integrals.

Example 35.1 The equation

$$y''+m^2y=0 \qquad (\S30, \text{Ex. } 30.2)$$

has, among others, the following integrating factors:

$$\cos mx, \qquad \sin mx, \qquad 2y'$$

which lead to the corresponding first integrals

$$y' \cos mx + my \sin mx = A,$$

$$y' \sin mx - my \cos mx = B,$$

$$y'^2 + m^2y^2 = c^2.$$

By eliminating y' between the first two we obtain the general integral

$$my = A \sin mx + B \cos mx.$$

The third may be obtained from the first two by squaring and adding; the relation between the constants is $A^2+B^2=c^2$.

§36 Problems involving curvature

The geometrical problems previously discussed (§§14, 15) depended upon a relationship between the slope y' and the coordinates (x, y) which was imposed at every point of a plane curve. Such problems thus led to differential equations of the first order. We now consider problems in which the curvature at every point obeys a specified law.

The radius of curvature at the point (x, y) is

$$\rho=\frac{(1+y'^2)^{3/2}}{y''};$$

the centre of curvature (ξ, η) relative to that point is given by the pair of equations

$$\xi=x-\frac{y'(1+y'^2)}{y''}, \qquad \eta=y+\frac{1+y'^2}{y''}.$$

The locus of the centre of curvature is the *evolute.* A whole family of parallel curves have the same evolute; they are its *involutes.*

As the above expressions involve y'', a problem that depends essentially upon curvature involves the integration of a differential equation of the second order.

As an example, we consider curves whose radius of curvature is proportional to the normal.

The length of the normal is understood to be the length of that portion intercepted between the curve and the x-axis, which is given by $y(1+y'^2)^{1/2}$. Thus if the normal is n times the radius of curvature, we have

$$y(1+y'^2)^{1/2}=\pm n(1+y'^2)^{3/2}/y''$$

or

$$yy''=\pm n(1+y'^2).$$

The ambiguous sign arises from the fact that each of the two lengths in question may be measured in either of two directions. Let us agree to measure them both away from the curve. If the curve were, at the point considered, above the x-axis, and concave upwards, both y and y'' would be positive, but the normal and radius of curvature would run in opposite directions. Hence our convention demands the negative sign.

The equation to which the problem has been reduced is therefore

$$yy'' = -n(1+y'^2).$$

Since x is absent, we write (§30)

$$y' = p, \qquad y'' = pp'(y)$$

and obtain

$$ypp'(y) + n(1+p^2) = 0.$$

Separating the variables and integrating, we obtain

$$1 + p^2 = (a/y)^{2n},$$

whence

$$x = \int \frac{\mathrm{d}y}{p} = \int \frac{\mathrm{d}y}{\sqrt{\{(a/y)^{2n} - 1\}}} + b$$

where a and b are the constants of integration.

We take two cases where the radius of curvature has the same direction as the normal ($n = 1, \frac{1}{2}$) and two cases in which the directions are opposed ($n = -1, -\frac{1}{2}$).

When $n = 1$,

$$x = \int y(a^2 - y^2)^{-1/2}\,\mathrm{d}y + b = -\sqrt{(a^2 - y^2)} + b$$

or

$$(x-b)^2 + y^2 = a^2.$$

Thus any circle with its centre on the x-axis satisfies the condition; in fact the normal and the radius are coincident.

When $n = \frac{1}{2}$,

$$x = \int \sqrt{\{y/(a-y)\}}\,\mathrm{d}y + b.$$

Writing $y = a\sin^2\theta$, $\mathrm{d}y = 2a\sin\theta\cos\theta$, we find that

$$x - b = 2a\int \sin^2\theta\,\mathrm{d}\theta = a\int (1 - \cos 2\theta)\,\mathrm{d}\theta = a(\theta - \tfrac{1}{2}\sin 2\theta)$$

or, if we put $2\theta = \phi$,

$$x = b + \tfrac{1}{2}a(\phi - \sin\phi), \qquad y = \tfrac{1}{2}a(1 - \cos\phi).$$

Thus the curve whose radius of curvature is twice its normal, and in the same direction as the normal, is a *cycloid*.

When $n = -1$,

$$x = \int a(y^2 - a^2)^{-1/2}\,\mathrm{d}y + b = a\cosh^{-1}(y/a) + b$$

or

$$y = a\cosh\left(\frac{x-b}{a}\right).$$

Thus when the radius of curvature equals the normal, but is oppositely directed, the curve is a *catenary*.

When $n = -\frac{1}{2}$,

$$x = \int (y/a - 1)^{-1/2}\,\mathrm{d}y + b = 2\sqrt{\{a(y-a)\}} + b$$

or

$$(x-b)^2 = 4a(y-a).$$

So when the radius of curvature is twice the normal, but is oppositely directed, the curve is a parabola with its axis perpendicular to the x-axis, and its latus rectum equal to four times the ordinate of its vertex.

5

Linear equations

§37 Form of the general integral

An equation of order n is said to be *linear* if it is linear in the dependent variable y and the derivatives $y', y'', \ldots, y^{(n)}$. Thus the most general linear equation of order n is of the form

$$p_0(x)y^{(n)}(x)+p_1(x)y^{(n-1)}(x)+p_{n-1}(x)y'+p_n(x)y=f(x). \tag{37.1}$$

If we denote by D the differential operator $\mathrm{d}/\mathrm{d}x$, we see that the left-hand member of this equation is derived from y by the application of the compound operator

$$p_0(x)D^n+p_1(x)D^{n-1}+\cdots+p_{n-1}(x)D+p_n(x),$$

and that if we write $\mathbf{L}$ for this operator the equation assumes the convenient abbreviated form

$$\mathbf{L}y=f(x). \tag{37.2}$$

We take it for granted that since the equation is of order n, its general integral depends on n distinct arbitrary constants, and proceed to consider the nature of this dependence.

Suppose that two distinct particular integrals of (37.2) are known, say $y_1(x)$ and $y_2(x)$. Then

$$\mathbf{L}y_1(x)=f(x), \qquad \mathbf{L}y_2(x)=f(x),$$

so that

$$\mathbf{L}y_2-\mathbf{L}y_1=0.$$

Now

$$p_{n-r}(x)D^r y_2-p_{n-r}(x)D^r y_1=p_{n-r}D^r(y_2-y_1)$$

and $\mathbf{L}(y_2-y_1)$ is merely a sum of terms like this, so that we have

$$\mathbf{L}(y_2-y_1)=\mathbf{L}y_2-\mathbf{L}y_1=0.$$

Thus if u represents the difference between any two solutions of (37.2), u

will satisfy the *reduced equation*

$$\mathbf{L}u=0, \tag{37.3}$$

which contains no term free from u or derivatives of u. For this reason we consider, first of all, the general solution of this reduced equation.

In the first place we observe that if C is any constant,

$$p_{n-r}(x)D^r(Cu)=Cp_{n-r}(x)D^r u,$$

and from the definition of the differential operator $\mathbf{L}$ we deduce that

$$\mathbf{L}(Cu)=C\mathbf{L}u.$$

So if u_1 is any solution of the reduced equation (37.3), we find that

$$\mathbf{L}(c_1u_1)=c_1\mathbf{L}u_1=0,$$

i.e. that c_1u_1 is a solution of the reduced equation for arbitrary constant c_1.

We next observe that if we have a number of particular integrals $u_1, u_2, \ldots, u_m$ then whatever the values assumed by the constants $c_1, c_2, \ldots, c_m$,

$$\mathbf{L}(c_1u_1+c_2u_2+\cdots+c_mu_m)=c_1\mathbf{L}u_1+c_2\mathbf{L}u_2+\cdots c_m\mathbf{L}u_m=0,$$

showing that $c_1u_1+c_2u_2+\cdots+c_mu_m$ is also an integral of the reduced equation.

Functions $u_1, u_2, \ldots, u_m$ are said to be *linearly independent* if it is impossible to find constants $c_1, c_2, \ldots, c_m$ *not all zero* such that

$$c_1u_1+c_2u_2+\cdots,+c_mu_m=0$$

identically. For example the three functions

$$\sin^2 x, \qquad \cos^2 x, \qquad \sin 2x$$

are linearly independent since no identity of the type

$$c_1 \sin^2 x+c_2 \cos^2 x+c_3 \sin 2x=0$$

exists. (To establish this result, suppose that such an identity does exist, then putting $x=0$ we find that $c_2=0$; then putting $x=\frac{1}{2}\pi$ we find that $c_1=0$ and hence that $c_3=0$). On the other hand the three functions

$$\sin^2 x, \qquad \cos^2 x \qquad \cos 2x$$

are linearly dependent since we have the identity

$$\sin^2 x-\cos^2 x+\cos 2x=0.$$

Suppose now that the functions $u_1, u_2, \ldots, u_n$ form a set of linearly independent solutions of the reduced equation (37.3), then the function u defined by

$$u = c_1 u_1 + c_2 u_2 + \cdots + c_n u_n,$$

with arbitrary constants $c_1, c_2, \ldots, c_n$ is also a solution of that equation, and since it is not identically zero for any choice of $c_1, c_2, \ldots, c_n$ (not all zero) and since, in addition, it contains the full complement of n arbitrary constants, it is the general solution of the reduced equation (37.3). Any such set $(u_1, u_2, \ldots, u_n)$ of linearly independent integrals is said to be a *fundamental* set of the equation (37.1).

We now set up a criterion by which we may determine whether or not a set of functions is linearly dependent. Suppose that $u_1, u_2, \ldots, u_n$ are linearly dependent; then there exist constants $c_1, c_2, \ldots, c_n$, not all zero, such that

$$c_1 u_1 + c_2 u_2 + \cdots + c_n u_n = 0.$$

Successively differentiating both sides of this identity we arrive at the set of relations

$$\begin{aligned} c_1 u_1' + c_2 u_2' + \cdots + c_n u_n' &= 0, \\ c_1 u_1'' + c_2 u_2'' + \cdots + c_n u_n'' &= 0, \\ &\vdots \\ c_1 u_1^{(n-1)} + c_2 u_2^{(n-1)} + \cdots + c_n u_n^{(n-1)} &= 0. \end{aligned}$$

Eliminating the constants $c_1, c_2, \ldots, c_n$ from this set of equations we deduce that a necessary condition for $u_1, u_2, \ldots, u_n$ to be linearly dependent is that

$$W(u_1, u_2, \ldots, u_n) = 0, \tag{37.4}$$

where the function W, called the *Wronskian* of the functions $u_1, u_2, \ldots, u_n$ is defined by the equation

$$W(u_1, \ldots, u_n) = \begin{vmatrix} u_1 & u_2 & \cdots & u_n \\ u_1' & u_2' & \cdots & u_n' \\ \vdots & \vdots & & \\ u_1^{(n-1)} & u_2^{(n-1)} & \cdots & u_n^{(n-1)} \end{vmatrix} \tag{37.5}$$

It can be shown that this condition is sufficient as well as necessary.

Example 37.1 Since

$$\begin{vmatrix} e^{ax} & e^{bx} & e^{cx} \\ a\,e^{ax} & b\,e^{bx} & c\,e^{cx} \\ a^2\,e^{ax} & b^2\,e^{bx} & c^2\,e^{cx} \end{vmatrix} = (b-c)(c-a)(a-b)\,e^{(a+b+c)x}$$

it follows that if a, b and c are unequal the functions e^{ax}, e^{bx}, e^{cx} are linearly independent.

Example 37.2 Since

$$\begin{vmatrix} e^{ax} & x\,e^{ax} & x^2\,e^{ax} \\ a\,e^{ax} & (1+ax)\,e^{ax} & (2x+ax^2)\,e^{ax} \\ a^2\,e^{ax} & (2a+a^2x)\,e^{ax} & (2+4ax+a^2x^2)\,e^{ax} \end{vmatrix} = 2\,e^{3ax},$$

we see that the functions

$$e^{ax}, \quad x\,e^{ax}, \quad x^2\,e^{ax},$$

are linearly independent.

We now return to the consideration of the original equation (37.1). Suppose that $y=v$ is any particular integral (involving no arbitrary constant) and $y=Y$ is the general integral. Then $u=Y-v$ must satisfy the reduced equation and must also contain n arbitrary constants; it will therefore be the general solution of the reduced equation. We therefore have

$$Y=c_1u_1+c_2u_2+\cdots+c_nu_n+v,$$

where $u_1, u_2, \ldots, u_n$ is a set of linearly independent solutions of the reduced equation (37.3). Thus the general solution of the inhomogeneous linear equation (37.1) has two components, namely

(i) the general solution of the reduced equation (37.3) involving n arbitrary constants, and known as the *complementary function*;
(ii) a particular integral of the full equation (and involving no arbitrary constant).

Example 37.3 To solve the equation

$$x(x^2+1)^2y''-(3x^2-1)(x^2+1)y'+4x(x^2-1)y=x^4-6x^2+1,$$

we notice that the reduced equation

$$x(x^2+1)^2u''-(3x-1)(x+1)u'+4x(x-1)u=0$$

has particular solutions

$$x^2+1, \qquad (x^2+1)\log x.$$

Since

$$\begin{vmatrix} x^2+1 & (x^2+1)\log x \\ 2x & 2x\log x + x + 1/x \end{vmatrix} = (x^2+1)^2/x,$$

these functions are linearly independent so that the general solution of the reduced equation is

$$u=(x^2+1)(c_1+c_2\log x).$$

Also, a particular integral of the original equation is $v(x)=x$; the general solution of the original equation is therefore

$$y=x+(x^2+1)(c_1+c_2\log x),$$

where c_1 and c_2 are arbitrary constants.

Example 37.4 The equation

$$x^2y''+2x(x-1)y'+(x^2-2x+2)y=x^3 \tag{37.6}$$

may be solved by noting that

$$x\,\mathrm{e}^{-x}, \qquad x^2\,\mathrm{e}^{-x}$$

are solutions of the reduced equation

$$x^2u''+2x(x-1)u'+(x^2-2x+2)u=0.$$

Also, since

$$\begin{vmatrix} x\,\mathrm{e}^{-x} & x^2\,\mathrm{e}^{-x} \\ (1-x)\,\mathrm{e}^{-x} & (2x-x^2)\,\mathrm{e}^{-x} \end{vmatrix} = x^2\,\mathrm{e}^{-2x},$$

these functions are linearly independent. The complementary function is therefore

$$u=(c_1+c_2x)x\,\mathrm{e}^{-x}.$$

Since $v(x)=x$ is a particular integral of (37.6) we deduce that

$$y=x+(c_1+c_2x)x\,\mathrm{e}^{-x},$$

with c_1 and c_2 arbitrary constants, is the general solution of (37.6).

§38 Depression of the order

When any integral of a reduced equation of order n is known, the equation can be transformed to another linear homogeneous equation of order $n-1$. Suppose that the known integral is $u_1(x)$. Then the transformation is

$$u(x)=u_1(x)V(x), \tag{38.1}$$

where V denotes the indefinite integral of an unknown function v; i.e. $V'=v$. This transformation effects the desired reduction in the case of general n, but, for simplicity, we shall illustrate its use only in the case $n=3$, in which case the relevant homogeneous equation of the third order is

$$p_0u'''+p_1u''+p_2u'+p_3u=0,$$

where the coefficients p_0, p_1, p_2 and p_3 are functions of x only. From (38.1) we deduce that

$$u'=u_1'V+u_1v, \qquad u''=u_1''V+2u_1'v+u_1v'$$

$$u'''=u_1'''V+3u_1''v+3u_1'v'+u_1v'',$$

where

$$V(x)=\int^x v(t)\,\mathrm{d}t.$$

Substituting in the original equation we obtain the equation

$$(p_0u'''+p_1u''+p_2u'+p_3u)V+P_2v+P_1v'+P_0v''=0,$$

where

$$P_0=p_0u_1, \qquad P_1=3p_0u_1'+p_1u_1, \qquad P_2=3p_0u''+2p_1u_1'+p_2u_1.$$

The coefficient of V vanishes, since u_1 is a solution of the original equation leaving the homogeneous linear equation of the second order

$$P_0v''+P_1v'+P_2v=0,$$

for the determination of the new dependent variable v.

Example 38.1 Since e^x is a particular integral of the equation

$$xy''-(2x+1)y'+(x+1)y=0, \tag{38.2}$$

we write

$$y=\mathrm{e}^x V, \quad y'=\mathrm{e}^x(V+v), \quad y''=\mathrm{e}^x(V+2v+v')$$

to obtain for v the first-order equation

$$xv' - v = 0.$$

Writing the solution of this equation as $v = 2c_1 x$, we see that $V = c_1 x^2 + c_2$, and hence that the solution of equation (38.2) is

$$y = e^x (c_1 x^2 + c_2).$$

§39 Reduction of a second-order equation to normal form

A linear differential equation of the second order is said to be in *normal form* when the term in y' is absent. In certain cases, by reducing the equation to normal form we can find its general solution and, in the theory of asymptotic solutions of differential equations there are distinct advantages in reducing an equation (of second order) to this form.

If we write the equation in the form

$$p_0(x)y'' + p_1(x)y' + p_2(x)y = 0, \tag{39.1}$$

and change the dependent variable from y to z, where $y = zu$, and use the results $y' = uz' + u'z$, $y'' = uz'' + 2u'z' + u''z$, we see that (39.1) is equivalent to

$$p_0 uz'' + (2p_0 u' + p_1 u)z' + (p_0 u'' + p_1 u' + p_2)z = 0.$$

So far we have said nothing about the choice of the function u; if we wish the coefficient of z' to be zero we must take u to be any solution of the equation $2p_0 u' + p_1 u = 0$. We must therefore take

$$u(x) = \exp\left\{-\frac{1}{2}\int \frac{p_1(x)}{p_0(x)}\,dx\right\} \tag{39.2}$$

In other words if we change the dependent variable from y to z by the transformation $y = uz$, where u is given by equation (39.2), the equation (39.1) is transformed to the normal form

$$z'' + Pz = 0, \tag{39.3}$$

with

$$P = \frac{p_0 u'' + p_1 u' + p_2}{p_0 u} \tag{39.4}$$

Example 39.1 If we take $p_0 = 1$, $p_1(x) = x^{-1}$ in (39.2) we find that

$u=x^{-1/2}$. Hence the substitution $y=x^{-1/2}z$ reduces Bessel's equation

$$\frac{d^2y}{dx^2}+\frac{1}{x}\frac{dy}{dx}+\left(1-\frac{\nu^2}{x^2}\right)y=0$$

to the standard form

$$\frac{d^2z}{dx^2}+\left(1-\frac{\nu^2-\frac{1}{4}}{x^2}\right)z=0.$$

Example 39.2 We can find the general solution of the equation

$$y''-4xy'+(4x^2-1)y=0,$$

by first reducing it to normal form.

Putting $y=uz$ where

$$u(x)=\exp\left(+\int 2x\,dx\right)=e^{x^2},$$

we find that

$$y'=e^{x^2}(z'+2xz), \qquad y''=e^{x^2}(z''+4xz'+2z+4x^2z)$$

and hence that z satisfies the equation

$$z''+z=0.$$

Since the general solution of this equation is

$$z=c_1\cos x+c_2\sin x$$

we see that the general solution of the original equation is

$$y=e^{x^2}(c_1\cos x+c_2\sin x).$$

§40 Reduced equation with constant coefficients

We now consider an equation of order n whose coefficients are constants. As no loss in generality results from taking the leading coefficient to be unity, the equation may be written as

$$(D^n+a_1D^{n-1}+\cdots+a_{n-1}D+a_n)u=0, \tag{40.1}$$

where $a_1, a_2, \ldots, a_n$ are constants, or in short as

$$F(D)u=0. \tag{40.2}$$

Now $F(D)$ is a polynomial, not in an algebraic symbol, but in the differential operator D; the laws of algebra cannot therefore be applied without prior justification. Our first step will be to show that $F(D)$ can be factorized in the same way, and with the same result as if D were an ordinary symbol in algebraic manipulations. Consider the product

$$(D-\alpha)(D-\beta)$$

in which α and β are constants, real or complex, and suppose that f is any twice-differentiable function. Then

$$\begin{aligned}(D-\alpha)(D-\beta)f &= (D-\alpha)(f'-\beta f)\\ &= f''-(\alpha+\beta)f'+\alpha\beta f\\ &= \{D^2-(\alpha+\beta)D+\alpha\beta\}f\end{aligned}$$

Similarly, we can show that

$$(D-\beta)(D-\alpha)f = \{D^2-(\alpha+\beta)D+\alpha\beta\}f$$

and hence that

$$(D-\beta)(D-\alpha) = (D-\alpha)(D-\beta).$$

Thus we have demonstrated that if α and β are constants $D-\alpha$ and $D-\beta$ commute. That they do not necessarily commute when α and β are not constants is easily shown by the following example:

$$(D-x)(D-1)f = (D-x)(f'-f) = f''-(x+1)f'+xf$$

$$(D-1)(D-x)f = (D-1)(f'-xf) = f''-(x+1)f'+(x-1)f$$

showing that

$$(D-x)(D-1) \neq (D-1)(D-x)$$

We have shown that any differential operator of the second order with constant coefficients may be factorized in the ordinary algebraic sense, into two permutable linear factors. By repeating the argument with n factors $D-\alpha_1, D-\alpha_2, \ldots, D-\alpha_n$ we can show that a differential operator of the nth order with constant coefficients can be decomposed into the produce of n such linear operators taken in any order.

Returning to the equation (40.1) we shall assume that the coefficients $a_1, a_2, \ldots, a_n$ are real so that the numbers $\alpha_1, \ldots, \alpha_n$ in the decomposition

$$F(D) = (D-\alpha_1)(D-\alpha_2)\cdots(D-\alpha_n) \tag{40.3}$$

are either real or paired as conjugate complex numbers.

Suppose that $D-\alpha_r$ is any linear factor of $F(D)$; then equation (40.2) may be written as

$$F_r(D)(D-\alpha_r)u=0.$$

Now

$$F(D)\,e^{\alpha_r x}=F_r(D)(D-\alpha_r)\,e^{\alpha_r x}=F_r(D)0=0$$

showing that $e^{\alpha_r x}$ is a solution of $F(D)=0$. Hence if the algebraic equation

$$F(m)=0 \tag{40.4}$$

has n distinct roots $\alpha_1, \alpha_2, \ldots, \alpha_n$, the functions

$$e^{\alpha_1 x}, e^{\alpha_2 x}, \ldots, e^{\alpha_n x}$$

are solutions of equation (40.2). Now the Wronskian of this set of functions is

$$\begin{vmatrix} 1 & 1 & \ldots & 1 \\ \alpha_1 & \alpha_2 & \ldots & \alpha_n \\ \alpha_1^2 & \alpha_2^2 & \ldots & \alpha_n^2 \\ \vdots & & & \\ \alpha_1^{n-1} & \alpha_2^{n-1} & \ldots & \alpha_n^{n-1} \end{vmatrix} \exp(\alpha_1 x+\alpha_2 x+\cdots+\alpha_n x_n)$$

or

$$\prod(\alpha_r-\alpha_s)\exp(\alpha_1 x+\alpha_2 x+\cdots+\alpha_n x_n)$$

where $\prod(\alpha_r-\alpha_s)$ denotes the product of all the differences $\alpha_r-\alpha_s$, $(r \neq s)$. Since $\alpha_1, \ldots, \alpha_n$ are distinct the product is non-zero as is the exponential factor, so the functions

$$e^{\alpha_1 x}, e^{\alpha_2 x}, \ldots, e^{\alpha_n x}$$

are linearly independent and form a fundamental set, provided that no two of $\alpha_1, \ldots, \alpha_n$ are equal. The case of repeated factors in $F(m)$ will be deferred to a later section (§42).

The form of fundamental set that has been obtained suggests that the most direct way of dealing with an equation of the form (40.2) is to use the result

$$F(D)\,e^{mx}=F(m)\,e^{mx} \tag{40.4a}$$

This shows that e^{mx} is a solution of $F(D)=0$ if m is a root of the *auxiliary equation* (40.4). By solving the auxiliary equation we obtain as many of the exponential members e^{mx} of the fundamental set as it has distinct roots, but when repeated roots are present, the solution is incomplete.

Example 40.1 Since the auxiliary equation of

$$y''' - 2y'' - 5y' + 6y = 0 \tag{40.5}$$

$m^3 - 2m^2 - 5m + 6 = 0$ has three distinct roots $-2, 1, 3$ it follows that the general solution of (40.5) is

$$y = c_1 \, e^{-2x} + c_2 \, e^{x} + c_3 \, e^{3x}$$

We now consider the situation when the auxiliary equation (40.4) has a pair of conjugate imaginary roots $\alpha + \beta i$ and $\alpha - \beta i$ so that we may write

$$F(D) = F_2(D)\{(D - \alpha)^2 + \beta^2\}$$

The root $\alpha + \beta i$ gives rise to a solution $A \, e^{(\alpha + \beta i)x}$ and the root $\alpha - \beta i$ to a solution $B \, e^{(\alpha - \beta i)x}$ so that the quadratic factor $(D - \alpha)^2 + \beta^2$ gives rise to a sum of terms

$$e^{\alpha x} \, [(A + B) \cos \beta x + (A - B) i \sin \beta x]$$

in the general solution. Writin $C_1 = A + B$, $C_2 = (A - B)i$ we see that the quadratic factor $m^2 - 2\alpha m + \alpha^2 + \beta^2$ in $F(m)$ gives rise to the term

$$e^{\alpha x} \, (C_1 \cos \beta x + C_2 \sin \beta x)$$

in the solution of $F(D)u = 0$.

Example 40.2 The auxiliary equation corresponding to

$$y'' + 6y' + 20y = 0$$

can be written

$$(m + 2)\{(m - 1)^2 + 9\} = 0,$$

so that the general solution is

$$y = c_1 \, e^{-2x} + (c_2 \cos 3x + c_3 \sin 3x) \, e^{x}.$$

Example 40.3 The auxiliary equation corresponding to

$$y^{(4)} - y = 0$$

can be written $(m - 1)(m + 1)(m^2 + 1) = 0$, so the general solution of the differential equation is

$$y = c_1 \, e^{x} + c_2 \, e^{-x} + c_3 \sin x + c_4 \cos x.$$

Example 40.4 The auxiliary equation corresponding to

$$y^{(4)}+4a^4y=0$$

can be written

$$(m^2-2ma+2a^2)(m^2+2ma+2a^2)=\{(m-a)^2+a^2\}\{(m+a)^2+a^2\}=0$$

and so has roots $a\pm ia$, $-a\pm ia$. The general solution is therefore

$$y=(c_1\cos ax+c_2\sin ax)\,\mathrm{e}^{ax}+(c_3\cos ax+c_4\sin ax)\,\mathrm{e}^{-ax};$$

this may also be written as

$$y=A\cos ax\cosh ax+B\sin ax\cosh ax+$$
$$+C\cos ax\sinh ax+E\sin ax\sinh ax.$$

§41 Properties of the operator $F(D)$

In what follows it will be necessary to apply a polynomial operator to a function of the form $\mathrm{e}^{ax}f(x)$, where f is a function of x, unspecified except for the stipulation that it shall be differentiable at least n times. Using Leibnitz's theorem for the rth derivative of a product, we have

$$\begin{aligned} D^r(\mathrm{e}^{ax}f) &= \sum_{m=0}^{r}\binom{r}{m}D^{r-m}f\,D^m(\mathrm{e}^{ax}) \\ &= \mathrm{e}^{ax}\sum_{m=0}^{r}\binom{r}{m}(a^mD^{r-m})f \\ &= \mathrm{e}^{ax}(D+a)^rf \end{aligned}$$

From this we deduce that if

$$F(D)=\sum_{r=0}^{n}c_rD^r,$$

then

$$F(D)\,\mathrm{e}^{ax}f(x)=\mathrm{e}^{ax}\sum_{r=0}^{n}c_r(D+a)^rf(x),$$

i.e.

$$F(D)\{\mathrm{e}^{ax}f(x)\}=\mathrm{e}^{ax}F(D+a)f(x) \tag{41.1}$$

or

$$F(D-a)\{\mathrm{e}^{ax}f(x)\}=\mathrm{e}^{ax}F(D)f(x). \tag{41.2}$$

At times it is necessary to operate on $\cos \omega x$ or $\sin \omega x$. We then have

$$D^{2m} \cos \omega x = (-1)^m \omega^{2m} \cos \omega x,$$

$$D^{2m} \sin \omega x = (-1)^m \omega^{2m} \sin \omega x,$$

$$D^{2m+1} \cos \omega x = (-1)^{m+1} \omega^{2m+1} \sin \omega x,$$

$$D^{2m+1} \sin \omega x = (-1)^m \omega^{2m+1} \cos \omega x.$$

Hence

$$\sum \lambda_m D^{2m} \cos \omega x = \sum \lambda_m (-\omega^2)^m \cos \omega x$$

$$\sum \mu_m D^{2m+1} \cos \omega x = -\omega \sum \mu_m (-\omega^2)^m \sin \omega x,$$

so if

$$F(D) = F_1(D^2) + DF_2(D^2) \tag{41.3}$$

then

$$F(D) \cos \omega x = F_1(-\omega^2) \cos \omega x - \omega F_2(-\omega^2) \sin \omega x \tag{41.4}$$

Similarly

$$F(D) \sin \omega x = F_1(-\omega^2) \sin \omega x + \omega F_2(-\omega^2) \cos \omega x. \tag{41.5}$$

From these equations we deduce immediately that

$$F(D)\left[\frac{F_1(-\omega^2) \cos \omega x + \omega F_2(-\omega^2) \sin \omega x}{F_1(-\omega^2) + \omega^2 F_2(-\omega^2)}\right] = \cos \omega x \tag{41.6}$$

and that

$$FD)\left[\frac{F_1(-\omega^2) \sin \omega x - \omega F_2(-\omega^2) \cos \omega x}{\{F_1(-\omega^2)\}^2 + \omega^2 \{F_2(-\omega^2)\}}\right] = \sin \omega x \tag{41.7}$$

provided that in both instances

$$\{F_1(-\omega^2)\}^2 + \omega^2 \{F_2(-\omega^2)\}^2 \neq 0. \tag{41.8}$$

The simplest case in which the condition (41.8) is not satisfied is when $F(D) = D^2 + \omega^2$. Since

$$D^2\{x \cos \omega x\} = -2\omega \sin \omega x - \omega^2 x \cos \omega x,$$

$$D^2\{x \sin \omega x\} = 2\omega \cos \omega x - \omega^2 x \sin \omega x$$

we deduce the pair of relations

$$\left.\begin{aligned} (D^2 + \omega^2)\{x \cos \omega x\} &= -2\omega \sin \omega x, \\ (D^2 + \omega^2)\{x \sin \omega x\} &= 2\omega \cos \omega x. \end{aligned}\right\} \tag{41.9}$$

§42 Repeated factors in the operator $F(D)$

A repeated factor in the polynomial $F(D)$ implies that two or more of the numbers $\alpha_1, \alpha_2, \ldots, \alpha_n$ are equal and hence that the number of linearly independent exponentials $e^{\alpha_1 x}, e^{\alpha_2 x}, \ldots, e^{\alpha_q x}$ is less than n. We shall now indicate how this deficiency is to be remedied.

Suppose that the factor $m-\alpha$ occurs p times in $F(D)$, so that the equation $F(D)u=0$ may be written

$$F_p(D)(D-\alpha)^p u=0, \tag{42.1}$$

where $F_p(D)$ is a polynomial in D of degree $n-p$. Then we look for a contribution to the general solution of the reduced equation $F(D)u=0$ arising from the general solution of the equation

$$(D-\alpha)^p u=0. \tag{42.2}$$

If we seek the general solution of this equation in the form

$$u(x)=e^{\alpha x}\, v(x) \tag{42.3}$$

and make use of the identity

$$(D-\alpha)^p(e^{\alpha x}\, v)=e^{\alpha x}\, D^p v,$$

which is the special case of (41.1) with $F(D)=(D-\alpha)^p$, we see that the substitution (42.3) reduces the equation (42.2) to the simple form

$$D^p v=0$$

whose general solution is any arbitrary polynomial of degree $p-1$ in x, say

$$v=c_1+c_2 x+\cdots+c_p x^{p-1}$$

where $c_1, c_2, \ldots, c_p$ denote arbitrary constants. Hence the contribution to the general solution of $F(D)u=0$ arising from the multiple factor $(D-\alpha)^p$ is

$$(c_1+c_2 x+\cdots+c_p x^{p-1})\, e^{\alpha x}. \tag{42.4}$$

When α is real, we may then take the arbitrary constants $c_1, c_2, \ldots, c_p$ to be real. Since the functions

$$e^{\alpha x}, x\, e^{\alpha x}, \ldots, x^{p-1}\, e^{\alpha x}$$

are linearly independent, the function (42.4) is the full contribution of the factor $(D-\alpha)^p$. The remaining terms of the complementary function are given by the general solution of the reduced equation

$$F_p(D)u=0. \tag{42.5}$$

Example 42.1 Since the auxiliary equation of

$$y''' - 3y' + 2y = 0 \qquad (42.6)$$

may be written $(m+2)(m-1)^2 y = 0$, we see that the factor $(m-1)^2$ gives the contribution $(c_1 + c_2 x)\,\mathrm{e}^x$ to the general solution of the given equation. In this case the equation corresponding to (42.5) is

$$(D+2)u = 0$$

with general solution $c_3\,\mathrm{e}^{-2x}$. Hence

$$y = (c_1 + c_2 x)\,\mathrm{e}^x + c_3\,\mathrm{e}^{-2x}$$

is the general solution of equation (42.6).

Example 42.2 The auxiliary equation of

$$y^{(4)} - 2y'' + y = 0 \qquad (42.7)$$

is $(m-1)^2(m+1)^2 = 0$. The factor $(m-1)^2$ gives rise to a pair of terms $(c_1 + c_2 x)\,\mathrm{e}^x$ in the general solution of (42.7) and the factor $(m+1)^2$ to a pair $(c_3 + c_4 x)\,\mathrm{e}^{-x}$, so that the general solution of (42.7) is

$$y = (c_1 + c_2 x)\,\mathrm{e}^x + (c_3 + c_4 x)\,\mathrm{e}^{-x}.$$

When α is complex, say $\alpha = \beta + i\gamma$, then we still adopt the solution

$$(A_1 + A_2 x + \cdots + A_p x^{p-1})\,\mathrm{e}^{\beta x + i\gamma x}$$

but with the constants $A_1, A_2, \ldots, A_p$ complex. Also, if the coefficients in the differential equation are all real, $F(D)$ has in addition to the factor $(D-\alpha)^p$ a second factor $(D-\bar{\alpha})^p$ where $\bar{\alpha} = \beta - i\gamma$ is the complex conjugate of α. This second factor yields a contribution

$$(A_{p+1} + A_{p+2}x + \cdots + A_{2p}x^{p-1})\,\mathrm{e}^{\beta x - i\gamma x},$$

where again $A_{p+1}, A_{p+2}, \ldots, A_{2p}$ denote complex constants. Adding the two contributions we obtain

$$\sum_{q=0}^{p-1} x^{q-1}\,\mathrm{e}^{\beta x}\,[A_q\,\mathrm{e}^{i\gamma x} + A_{p+q}\,\mathrm{e}^{-i\gamma x}].$$

The real part of this will also be a solution, so taking the real part and writing

$$b_q = \mathrm{Re}\,(A_{p+q} + A_p), \qquad c_q = \mathrm{Im}\,(A_{p+q} - A_p),$$

we find that the contribution to the general solution of $F(D)u = 0$ arising from a factor

$$\{(m-\beta)^2+\gamma^2\}^p$$

in $F(m)$ is

$$\sum_{q=0}^{p-1}(b_q \cos \gamma x + c_q \sin \gamma x)x^{q-1}\,\mathrm{e}^{\beta x}.$$

Example 42.3 Since the auxiliary equation of the sixth-order equation

$$y^{(6)}+a^2y^{(4)}-a^4y''-a^6y=0$$

can be factorized to yield $(m^2-a^2)(m^2+a^2)^2=0$, we see that

$$y=c_1\,\mathrm{e}^{ax}+c_2\,\mathrm{e}^{-ax}+(c_3+c_4x)\cos ax+(c_5+c_6x)\sin ax$$

is the general solution of the equation.

§43 Use of the Laplace transform

The solution of equations of the type $F(D)u=0$, where

$$F(D)=D^n+a_1D^{n-1}+\cdots+a_n$$

can be obtained by means of the Laplace transform, whose main properties are described in §A1. The success of the method arises from the fact that if $\bar{u}=\mathscr{L}u$ is the Laplace transform of u,

$$\mathscr{L}[D^m u(x); p]=p^m\bar{u}(p)-\sum_{r=0}^{m-1}p^{m-r-1}u^{(r)}(0)$$

(cf. equation (A1.10)). From this it follows immediately that

$$\mathscr{L}[F(D)u(x); p]=F(p)\bar{u}(p)-G(p) \tag{43.1}$$

where

$$G(p)=\sum_{r=0}^{n-1}p^{n-r-1}u^{(r)}(0)+\sum_{m=1}^{n-1}a_{n-m}\sum_{r=0}^{m-1}p^{m-r-1}u^{(r)}(0). \tag{43.2}$$

If the values of $u(0), u'(0), \ldots, u^{(n-1)}(0)$ are assigned, then we can easily calculate the coefficients of the polynomial $G(p)$. We can rewrite

$$G(p)=\sum_{r=0}^{n-1}c_r p^{n-r-1} \tag{43.3}$$

where $c_0 = u(0)$, $c_1 = u'(0) + a_1 u(0)$ and

$$c_r = \sum_{m=0}^{r=1} a_m u^{(r-1-m)}(0), \qquad (2 < r < n-1) \tag{43.4}$$

with the convention that $a_0 = 1$.

Taking the Laplace transform of both sides of the equation $F(D)u = 0$ we find that $\bar{u} = \mathscr{L}u$ satisfies the equation

$$F(p)\bar{u}(p) = G(p)$$

so that

$$u(x) = \mathscr{L}^{-1}\left[\frac{G(p)}{F(p)}; x\right]. \tag{43.5}$$

When the solution of an initial-value problem is required the form (43.2) is the one which more quickly yields the desired result. On the other hand if we wish to find the general solution of a differential we would use the form (43.3). However, in solving problems, it is more important to use the appropriate method than it is to attempt to remember the formulae (43.2), (43.3).

Example 43.1 To find the solution of the initial-value problem

$$y''' - y'' + y' - y = 0, \qquad y(0) = 0, \qquad y'(0) = 1, \qquad y''(0) = -1$$

we make use of the results

$$\mathscr{L}[y'; p] = p\bar{y} - y(0) = p\bar{y}$$

$$\mathscr{L}[y''; p] = p^2\bar{y} - py(0) - y'(0) = p^2\bar{y} - 1$$

$$\mathscr{L}[y'''; p] = p^3 y - p^2 y(0) - py'(0) = p^3\bar{y} - p + 1$$

where $\bar{y} = \mathscr{L}y$, to obtain the equation

$$(p^3 - p^2 + p - 1)\bar{y}(p) = p - 2.$$

This has solution

$$\bar{y}(p) = \frac{p-2}{(p-1)(p^2+1)}.$$

Resolving into partial fractions the expression of the right-hand side of this equation we can rewrite it is

$$\bar{y}(p) = \frac{1}{2} \cdot \frac{p}{p^2+1} + \frac{3}{2} \cdot \frac{1}{p^2+1} - \frac{1}{2} \cdot \frac{1}{p-1}.$$

Taking inverse Laplace transforms of both sides of this equation we obtain

$$y(x) = \tfrac{1}{2}\cos x + \tfrac{3}{2}\sin x - \tfrac{1}{2}e^x$$

as the solution of the stated initial-value problem.

Example 43.2 To find the general solution of the equation

$$y''' + y'' - y' - y = 0$$

we use the results quoted at the beginning of Example *43.1* to obtain

$$\begin{aligned}(p^3+p^2-p-1)\bar{y}(p) &= p^2 y(0) + p\{y'(0)+y(0)\} + y''(0) + y'(0) - y(0) = \\ &= \{\tfrac{1}{2}y(0) + \tfrac{1}{4}y'(0) + \tfrac{1}{4}y''(0)\}(p+1)^2 + \\ &\quad + \{\tfrac{1}{2}y(0) - \tfrac{1}{4}y'(0) - \tfrac{1}{4}y''(0)\}(p^2-1) + \\ &\quad + \{y(0) - \tfrac{1}{2}y'(0) - \tfrac{1}{2}y''(0)\}(p-1) = \\ &= c_1(p+1)^2 + c_2(p^2-1) + c_3(p-1),\end{aligned}$$

where c_1, c_2, c_3 are linearly independent constants, since $y(0)$, $y'(0)$, $y''(0)$ are, in the general solution, linearly independent. Since, in this case,

$$F(p) = (p+1)(p^2-1) = (p+1)^2(p-1),$$

we see that

$$\bar{y}(p) = \frac{c_1}{p-1} + \frac{c_2}{p+1} + \frac{c_3}{(p+1)^2}.$$

Taking inverse Laplace transforms we obtain the solution

$$y(x) = c_1 e^x + (c_2 + c_3 x)e^{-x}.$$

It is obvious from this example that the use of the Laplace transform to derive the complementary function of a linear differential equation is more cumbersome than the use of the methods developed in §§41, 42. Where the use of the Laplace transform does effect an obvious economy is in deriving a particular integral $v(x)$ of the equation

$$F(D)y = f(x) \tag{43.6}$$

Since a particular integral is any solution of the equation, we can take $v(x)$ to be the solution of the initial-value problem

$$F(D)v = f(x), \qquad v(0) = v'(0) = \cdots = v^{(n-1)}(0) = 0.$$

Taking into account the initial values we deduce from equation (A1.10) and the definition of $F(D)$ – cf. equations (40.1), (40.2) – that

$$F(p)\bar{v}(p)=\bar{f}(p)$$

where $\bar{v}=\mathscr{L}v$, $\bar{f}=\mathscr{L}f$. Solving this equation for $\bar{v}(p)$ we find that

$$\bar{v}(p)=\bar{K}(p)\bar{f}(p),$$

where $\bar{K}(p)=1/F(p)$. If a function K exists such that

$$\bar{K}(p)=\frac{1}{F(p)},$$

i.e. if $K(x)=\mathscr{L}^{-1}[1/F(p); x]$, then making use of the convolution theorem expressed by equation (A1.16) we see that equation (43.6) yields the particular integral

$$v(x)=\int_0^x f(t)K(x-t)\,\mathrm{d}t. \tag{43.7}$$

Example 43.3 For the equation

$$y''+\omega^2 y=f(x), \tag{43.8}$$

$F(p)=p^2+\omega^2$, so that

$$\bar{K}(p)=\frac{1}{p^2+\omega^2};$$

hence

$$v(x)=\frac{1}{\omega}\int_0^x f(t)\sin\{\omega(x-t)\}\,\mathrm{d}t$$

is a particular integral of equation (43.8). The general solution of the equation (43.8) is therefore

$$y(x)=c_1\cos\omega x+c_2\sin\omega x+\frac{1}{\omega}\int_0^x f(t)\sin\{\omega(x-t)\}\,\mathrm{d}t. \tag{43.9}$$

Although general formulae of the kind (43.8) and (43.9) are useful, it is often preferable to use the *method* of deriving the formulae than to use the formulae themselves.

Example 43.4 To solve the initial-value problem

$$y''+\omega^2 y=\sin(\omega x),\qquad y(0)=0,\quad y'(0)=1$$

we take the Laplace transform of both sides of the equation to obtain the equation

$$p^2\bar{y}(p)-1+\omega^2\bar{y}(p)=\frac{\omega}{p^2+\omega^2}$$

for $\bar{y}=\mathscr{L}y$. Writing this equation as

$$\bar{y}(p)=\frac{1}{p^2+\omega^2}+\frac{\omega}{(p^2+\omega^2)^2}$$

we see from entries (11), (14) in the table of Laplace transforms on p. 186 that the required solution is

$$y(x)=\frac{1}{\omega}\sin\omega x+\frac{1}{2\omega^2}\{\sin\omega x-\omega x\cos\omega x\}. \qquad (43.10)$$

Although the primary use of the Laplace transform is in the solution of linear equations with constant coefficients, there are certain second-order equations with variable coefficients which can be solved by the transform method. In problems of this kind the transform of the desired solution is determined by a first-order differential equation. The key relation is equation (A1.12). We shall illustrate the method by considering a specific example.

Example 43.5 We shall find the solution of the equation

$$xy''+y'+xy=0 \qquad (43.11)$$

(*Bessel's equation* of order zero) satisfying the initial condition $y(0)=1$ and the additional condition

$$\lim_{x\to\infty} y(x)=0. \qquad (43.12)$$

If $\bar{y}=\mathscr{L}y$, then since $y(0)=1$,

$$\mathscr{L}[y';p]=p\bar{y}(p)-1$$

and

$$\mathscr{L}[y'';p]=p^2\bar{y}(p)-p-k,$$

where $k=y'(0)$ is unknown. From equation (A1.12) we then deduce that

$$\mathscr{L}[xy(x);p]=-\bar{y}'(p)$$

$$\mathscr{L}[xy''(x);p]=-p^2\bar{y}'(p)-2p\bar{y}(p)+1.$$

Hence the Laplace transform of (43.11) is

$$(p^2+1)\bar{y}'(p)+p\bar{y}(p)=0.$$

The solution of this equation is

$$\bar{y}(p) = C(p^2+1)^{-1/2},$$

where C is a constant. The condition $y(0)=1$ taken with equation (A1.18) gives

$$1 = y(0) = \lim_{p\to\infty} p\bar{y}(p) = C,$$

so that

$$\bar{y}(p) = (p^2+1)^{-1/2}. \qquad (43.13)$$

Also, using equation (A1.19), we see that the solution $y(x) = \mathscr{L}^{-1}[(p^2+1)^{-1/2}; x]$ satisfies the condition (43.12).

We shall now derive an infinite series for this solution.

Writing $\bar{y}(p) = p^{-1}(1+p^{-2})^{-1/2}$ and using the binomial theorem we obtain the series expansion

$$\bar{y}(p) = \sum_{n=0}^{\infty} \frac{(\frac{1}{2})_n}{n!} \frac{(-1)^n}{p^{2n+1}},$$

where

$$\left(\frac{1}{2}\right)_n = \frac{1}{2}\cdot\frac{3}{2}\cdot\frac{5}{2}\cdot\ldots\cdot\frac{2n-1}{2} = \frac{(2n)!}{2^{2n}n!}.$$

Hence

$$\bar{y}(p) = \sum_{n=0}^{\infty} \frac{(-1)^n}{n!\,n!}\left(\frac{1}{2}\right)^{2n} \frac{(2n)!}{p^{2n+1}}.$$

Applying the inverse Laplace transform to each side of this equation and using entry (2) in the table of Laplace transforms, we find that $y(x) = J_0(x)$, where

$$J_0(x) = \sum_{n=0}^{\infty} \frac{(-1)^n(\frac{1}{2}x)^{2n}}{n!\,n!}. \qquad (43.14)$$

The function $J_0(x)$ defined by equation (43.14) is called the *Bessel function of the first kind of order zero.*

§44 Inverse operators

Looking at the problem symbolically we could say that a particular integral of the equation

$$F(D)y = f(x) \qquad (44.1)$$

is

$$y=[F(D)]^{-1}f(x). \tag{44.2}$$

The problem then is to define precisely what is meant by $[F(D)]^{-1}$, the inverse of $F(D)$, and then to establish properties of the inverse operator.

Difficulties arise in even the simplest case. For instance the differential equation

$$Dy=f(x)$$

has solution

$$y=\int f(x)\,dx$$

so that we might identify the operator D^{-1} with $\int$, the operator of indefinite integration. This interpretation is justified by the identity

$$D(D^{-1}f)=f$$

showing that

$$DD^{-1}=I \tag{44.3}$$

where I denotes the identity operator. On the other hand $D^{-1}D$ is not unique. For example, if we take $f(x)=x^2$

$$D^{-1}Df=D^{-1}(2x)=x^2+c=f+c,$$

where c is an arbitrary constant. Since, in general, c is non-zero, we see that, in general,

$$D^{-1}D\neq I. \tag{44.4}$$

In the strictly algebraic sense we can therefore say that the differential operator D does not possess a unique inverse D^{-1}. All that equation (44.3) shows is that the operator D has a *unique right inverse* D^{-1}; similarly equation (44.4) shows that D does **not** possess a *unique left inverse.*

Just as D^r indicates the operation of r-fold differentiation, so D^{-r} will denote r-fold integration. This is a process which has a clear meaning, and the relation

$$D^r(D^{-r})=I$$

is obviously true. On the other hand

$$D^{-r}(D^r)=I$$

only if we make a suitable choice of the coefficients in the polynomial of degree $r-1$ introduced by the integration.

The operator $[F(D)]^{-1}$ inverse to the polynomial differential operator $F(D)$ has not yet been identified in the general case with any manipulative process, and can be interpreted only in special cases which will now be considered.

From the relation (40.4) we deduce that

$$[F(D)]^{-1}\,e^{ax}=\frac{e^{ax}}{F(a)},\qquad F(a)\neq 0. \tag{44.5}$$

It remains to consider the case in which $F(a)=0$. We recall the relation (41.1)

$$F(D)\{e^{ax}\,g(x)\}=e^{ax}\,F(D+a)g(x).$$

Now $F(D+a)g$ is a function of x which we may denote by $f(x)$, i.e. $F(D+a)g=f$, so that we may take

$$g=[F(D+a)]^{-1}\,f(x)$$

in this last equation to obtain the identity

$$F(D)\{e^{ax}\,[F(D+a)]^{-1}\,f(x)\}=e^{ax}\,f(x).$$

This in turn can be written in the inverse form

$$[F(D)]^{-1}\{e^{ax}\,f(x)\}=e^{ax}\,[F(D+a)]^{-1}f(x), \tag{44.6}$$

a particular case of which,

$$(D-a)^{-r}\,e^{ax}\,f(x)=e^{ax}\,D^{-r}f(x), \tag{44.7}$$

should be noticed. In the special case in which $f(x)$ is the unit function this last equation reduces to

$$(D-a)^{-r}\,e^{ax}=e^{ax}\,D^{-r}1;$$

noticing that by ordinary integration

$$D^{-r}1=\frac{x^r}{r!}$$

(ignoring the arbitrary constant that enters at each step) we see that this is equivalent to

$$(D-a)^{-r}\,e^{ax}=x^r\,e^{ax}/r! \tag{44.8}$$

In general, if we can factorize F in the form

$$F(D)=F_1(D)F_2(D),$$

then

$$[F(D)]^{-1} = [F_2(D)]^{-1}[F_1(D)]^{-1}$$

(the reversal rule). Hence, if

$$F(D) = G(D)(D-a)^r, \qquad G(a) \neq 0$$

then

$$[F(D)]^{-1} e^{ax} = (D-a)^{-r}[G(D)]^{-1} e^{ax}$$

and, since $G(a) \neq 0$, the right-hand side of this equation becomes

$$(D-a)^{-r} e^{ax}/G(a).$$

Using equation (44.8) we therefore obtain the result

$$[G(D)(D-a)^r]^{-1} = \frac{x^r e^{ax}}{r!\, G(a)}, \tag{44.9}$$

which generalizes (44.5).

Thus we have shown that

(i) a particular integral of $F(D)y = e^{ax}$ is $y = e^{ax}/F(a)$, provided $F(a) \neq 0$;
(ii) when $F(D)$ is of the form $G(D)(D-a)^r$, where $G(a) \neq 0$, i.e. when the auxiliary equation has a as an r-fold root, the simplest particular integral of $F(D)y = e^{ax}$ is $x^r e^{ax}/\{r!\, G(a)\}$.

Example 44.1 Since the auxiliary equation of

$$y'' + y' - 6y = 8 e^{3x} \tag{44.10}$$

has roots 2 and -3, the complementary function is

$$u = c_1 e^{2x} + c_2 e^{-3x}$$

This does not contain a term in e^{3x}, so that a particular integral is

$$\frac{8}{3^2 + 3 - 6} e^{3x} = \tfrac{4}{3} e^{3x}$$

and the general solution is

$$y = \tfrac{4}{3} e^{3x} + c_1 e^{2x} + c_2 e^{-3x} \tag{44.11}$$

Note. When a particular integral satisfying certain specified conditions is required, it may be deduced from the general solution by assigning the correct values to the constants introduced through the complementary function. For example, to determine the solution of equation (44.10) satisfying the initial conditions $y(0)=0, y'(0)=3$ we must choose the constants c_1, c_2 in the solution (44.11) such that

$$0=\tfrac{4}{3}+c_1+c_2, \qquad 3=4+2c_1-3c_2.$$

Solving these equations we find that $c_1=-1$ and $c_2=-\tfrac{1}{3}$, so that the required solution is

$$y=\tfrac{4}{3}\,\mathrm{e}^{3x}-\mathrm{e}^{2x}-\tfrac{1}{3}\,\mathrm{e}^{-3x}.$$

An obvious alternative method of solution of this initial-value problem is that of the Laplace transform (which the reader should attempt for himself).

Example 44.2 If we wish to solve the initial-value problem

$$y''-3y'+2y=\mathrm{e}^x+\mathrm{e}^{2x}, \qquad y(0)=y'(0)=1,$$

we note that the particular integral may be written

$$\begin{aligned} y&=(D-1)^{-1}(D-2)^{-1}\,\mathrm{e}^x+(D-2)^{-1}(D-1)^{-1}\,\mathrm{e}^{2x}=\\ &=-(D-1)^{-1}\,\mathrm{e}^x+(D-2)^{-1}\,\mathrm{e}^{2x}=\\ &=-x\,\mathrm{e}^x+x\,\mathrm{e}^{2x}. \end{aligned}$$

Hence the required solution is

$$y=(c_1-x)\,\mathrm{e}^x+(c_2+x)\,\mathrm{e}^{2x},$$

where the constants c_1, c_2 are chosen to be such that

$$1=c_1+c_2, \qquad 1=c_1+2c_2.$$

The solution of these equations is $c_1=1$, $c_2=0$, so that the required solution is

$$y=(1-x)\,\mathrm{e}^x+x\,\mathrm{e}^{2x}.$$

We have similar results for the effect of some inverse operators on circular functions. For example, from equations (41.4) and (41.5) we deduce immediately that if

$$F(D)=F_1(D^2)+DF_2(D^2) \tag{44.12}$$

then

$$F(D)[A(\omega)\cos\omega x+B(\omega)\sin\omega x]=\cos\omega x$$
$$F(D)[A(\omega)\sin\omega x-B(\omega)\cos\omega x]=\sin\omega x$$

where

$$A(\omega)=\frac{F_1(-\omega^2)}{\{F_1(-\omega^2)\}^2+\omega^2\{F_2(-\omega^2)\}^2},$$

$$B(\omega)=\frac{\omega F_2(-\omega^2)}{\{F_1(-\omega^2)\}^2+\omega^2\{F_2(-\omega^2)\}^2}. \qquad (44.13)$$

Provided the denominations of these expressions do not vanish, we may write this pair of relations in the inverse form

$$[F(D)]^{-1}\cos\omega x=A(\omega)\cos\omega x+B(\omega)\sin\omega x$$

$$[F(D)]^{-1}\sin\omega x=A(\omega)\sin\omega x-B(\omega)\cos\omega x \qquad (44.14)$$

with A, B defined by equations (44.13).

Some special cases are noteworthy. Since

$$(D^2+\omega^2)(x\cos\omega x)=-2\omega\sin\omega x$$

$$(D^2+\omega^2)(x\sin\omega x)=2\omega\cos\omega x$$

we see that we can write

$$[D^2+\omega^2]^{-1}\sin\omega x=-\frac{1}{2\omega}x\cos\omega x,$$

$$(44.15)$$

$$[D^2+\omega^2]^{-1}\cos\omega x=\frac{1}{2\omega}x\sin\omega x.$$

Alternatively, sines and cosines may be replaced by imaginary exponentials. We illustrate this procedure by:

Example 44.3 To find a particular integral of the equation

$$y''+\omega^2 y=e^{ax}\cos\omega x$$

we use equation (44.5) to obtain

$$(D^2+\omega^2)^{-1}e^{(a+i\omega)x}=\{(a+i\omega)^2+\omega^2\}^{-1}e^{(a+i\omega)x}=$$

$$=\{a(a+2i\omega)\}^{-1}e^{(a+i\omega)x}=$$

$$=\frac{1-2i\omega/a}{a^2+4\omega^2}e^{(a+i\omega)x}$$

Taking real parts of both sides of this relation we obtain the particular integral

$$\frac{1}{a^2+4\omega^2}\left[\cos\omega x+\frac{2\omega}{a}\sin\omega x\right]e^{ax}$$

On the other hand, if we had taken imaginary parts we should have obtained

$$\frac{1}{a^2+4\omega^2}\left[\sin\omega x-\frac{2\omega}{a}\cos\omega x\right]e^{ax}$$

as a particular integral of the equation

$$y''+\omega^2 y=e^{ax}\sin\omega x.$$

Example 44.4 If j denotes the current intensity at time t in a circuit containing an inductance L, a resistance R and a capacity C, then

$$L\frac{d^2j}{dt^2}+R\frac{dj}{dt}+\frac{j}{C}=\frac{dE}{dt},$$

where E is the impressed electromotive force.

The auxiliary equation is $CLm^2+CRm+1=0$ so that a particular solution of the reduced equation is e^{rt}, where

$$r=-\frac{R}{2L}\left[1\pm\left(1-\frac{4L}{R^2C}\right)^{1/2}\right]$$

and three cases are possible:

(i) If $R^2C>4L$, the radical is real but numerically less than unity; hence the complementary function is of the form $c_1\,e^{-at}+c_2\,e^{-bt}$, with a and b both real and positive.

(ii) If $R^2C=4L$, the complementary function has the form $(c_1+c_2t)\,e^{-at}$, where $a=\frac{1}{2}(R/L)$.

(iii) If $R^2C>4L$ the radical is imaginary and the complementary function takes the form

$$e^{-at}(c_1\cos bt+c_2\sin bt),\qquad (a>0).$$

In all three cases, the exponential becomes small very rapidly, so that in practice is it negligible, so that the complementary function represents a *transient* effect.

The most important form of E consists of a sum of sinusoidal terms; as their effect is additive we may confine our attention to a single term $e=$

$E_0 \sin(\omega t + \varepsilon)$. Since

$$(CLD^2 + CRD + 1)^{-1} D\, e^{i(\omega t+\varepsilon)} = \frac{i\omega}{1 - CL\omega^2 + iCR\omega}\, e^{i(\omega t+\varepsilon)} =$$

$$= \frac{\omega}{A} \exp\{i(\omega t + \varepsilon + \tfrac{1}{2}\pi - \beta)\}$$

where

$$A = \{(1 - CL\omega^2)^2 + C^2R^2\omega^2\}^{1/2},$$

$$\beta = \tan^{-1}\{CR\omega/(1 - CL\omega^2)\},$$

we find on taking imaginary parts that a suitable particular integral is

$$j = (\omega C E_0/A) \sin(\omega t + \varepsilon + \tfrac{1}{2}\pi - \beta).$$

This gives the *steady-state* current whose phase lags behind that of the electromotive force by $\frac{1}{2}\pi - \beta$.

We now consider the series development of a differential operator $F(D)$ with constant coefficients. Suppose that

$$F(m) = (m - \alpha)^{r_1}(m - \beta)^{r_2} \cdots$$

then we can resolve $1/F(m)$ into a sum of partial fractions of the form

$$\frac{1}{F(m)} = \frac{A_1}{m - \alpha} + \frac{A_2}{(m - \alpha)^2} + \cdots + \frac{A_r}{(m - \alpha)^{r_1}} + \frac{B_1}{m - \beta} + \frac{B_2}{(m - \beta)^2} + \cdots$$

This algebraic formula encourages us to look at a formula of the type

$$[F(D)]^{-1} = A_1(D - \alpha)^{-1} + \cdots + A_r(D - \alpha)^{-r} + B_1(D - \beta)^{-1} + \cdot \quad (44.16)$$

as yielding an interpretation of the inverse of the differential operator. This involves us in looking at the meaning of $(D - \alpha)^{-s}$ etc. and then verifying that, defining $[F(D)]^{-1}$ by (44.16), $F(D)[F(D)]^{-1}$ is the identity operator.

We may now interpret $(D - \alpha)^{-1}x^n$, where n is a positive integer as

$$-\frac{1}{\alpha}\left(1 - \frac{D}{\alpha}\right)^{-1} x^n = -\frac{1}{\alpha}\sum_{s=0}^{n} \alpha^{-s} D^s x^n,$$

the series terminating in D^n since $D^s x^n = 0$, $(s > n)$. Using the formula

$$D^s x^n = \frac{n!}{(n - s)!}\, x^{n-s}$$

we arrive at the formula

$$(D-\alpha)^{-1}x^n = -\frac{1}{\alpha}\sum_{s=0}^{n}\frac{n!}{(n-s)!}\frac{x^{n-s}}{\alpha^s}. \tag{44.17}$$

From (44.17) we obtain the relations

$$D(D-\alpha)^{-1}x^n = -\sum_{s=0}^{n-1}\frac{n!}{(n-s-1)!}\frac{x^{n-s-1}}{\alpha^{s+1}}$$

$$\alpha(D-\alpha)^{-1}x^n = -x^n - \sum_{s=1}^{n}\frac{n!}{(n-s)!}\frac{x^{n-s}}{\alpha^s} =$$

$$= -x^n - \sum_{s=0}^{n-1}\frac{n!\,x^{n-s-1}}{(n-s-1)!\,\alpha^{s+1}}$$

immediately and from them we verify that the formula (44.17) yields the necessary condition

$$(D-\alpha)\{(D-\alpha)^{-1}x^n\} = x^n.$$

In similar fashion we can interpret $(D-\alpha)^{-2}x^n$ as

$$\frac{1}{\alpha^2}\left(1-\frac{D}{\alpha}\right)^{-2}x^n = \frac{1}{\alpha^2}\sum_{s=0}^{n}\frac{(s+1)}{\alpha^s}D^s x^n$$

and expanding each term in (44.16) we arrive at the following interpretation of $[F(D)]^{-1}x^n$:

$$[F(D)]^{-1}x^n = (c_0 + c_1 D + \cdots + c_n D^n)x^n, \tag{44.18}$$

where the series on the right-hand side of this equation contains the first $n+1$ terms of the algebraic development of $[F(D)]^{-1}$ in ascending powers of D. It terminates in D^n since $D^r x^n = 0$ for $r > n$. It follows that if $p_n(x)$ is a polynomial of degree n,

$$[F(D)]^{-1}p_n(x) = (c_0 + c_1 D + c_2 D^2 + \cdots + c_n D^n)p_n(x). \tag{44.19}$$

Example 44.5 We may write a particular integral of the equation

$$y'' - 7y' + 12y = e^{2x}(x^3 - 5x^2)$$

$$y = (3-D)^{-1}(4-D)^{-1}e^{2x}(x^3 - 5x^2) =$$

$$= e^{2x}(1-D)^{-1}(2-D)^{-1}(x^3 - 5x^2).$$

Now retaining terms up to D^3 we have

$$\begin{aligned}(1-D)^{-1}(2-D)^{-1} &= \tfrac{1}{2}(1-D)^{-1}(1-\tfrac{1}{2}D)^{-1}= \\ &= \tfrac{1}{2}(1+D+D^2+D^3)(1+\tfrac{1}{2}D+\tfrac{1}{4}D^2+\tfrac{1}{8}D^3)= \\ &= \tfrac{1}{2}+\tfrac{3}{4}D+\tfrac{7}{8}D^2+\tfrac{15}{16}D^3\end{aligned}$$

so the required particular integral is

$$\begin{aligned}&e^{2x}\{\tfrac{1}{2}(x^3-5x^2)+\tfrac{3}{4}(3x^2-10x)+\tfrac{7}{8}(6x-10)+\tfrac{15}{16}\cdot 6\}= \\ &\qquad = e^{2x}(\tfrac{1}{2}x^3-\tfrac{1}{4}x^2-\tfrac{9}{4}x-\tfrac{25}{8}).\end{aligned}$$

§45 General solution by quadratures

When the roots of the auxiliary equation $F(m)=0$ are simple, the equation $F(D)y=f(x)$ may be integrated completely by quadratures. If

$$F(m)=(m-\alpha_1)(m-\alpha_2)\cdots(m-\alpha_n),$$

with $\alpha_1, \alpha_2, \ldots, \alpha_n$ distinct, then

$$\frac{1}{F(m)}=\sum_{r=1}^{n}\frac{1}{F'(\alpha_r)(m-\alpha_r)}$$

This suggests the possible formula

$$[F(D)]^{-1}f(x)=\sum_{r=1}^{n}\frac{1}{F'(\alpha_r)}(D-\alpha_r)^{-1}f(x). \tag{45.1}$$

Now $(D-\alpha_r)^{-1}f(x)$ is the solution of the differential equation $y'-\alpha_r y=f$ and this is readily seen to be

$$c_r e^{\alpha_r x}+e^{\alpha_r x}\int e^{-\alpha_r x}f(x)\,dx \tag{45.2}$$

or

$$e^{\alpha_r x}\int_{c_r}^{x} e^{-\alpha_r t}f(t)\,dt \tag{45.3}$$

where the lower terminal c_r is constant. From (45.1) and (45.2) we deduce that the general solution of the differential equation

$$F(D)y=f(x) \tag{45.4}$$

when the zeros of $F(m)$ are simple is

$$y(x)=u(x)+\sum_{r=1}^{n}\frac{1}{F'(\alpha_r)}e^{\alpha_r x}\int e^{-\alpha_r x}f(x)\,dx$$

where

$$u(x) = \sum_{r=1}^{n} c_r \, e^{\alpha_r x}$$

is the relevant complementary function. On the other hand, if we use (45.1) and (45.3) we obtain

$$y(x) = \sum_{r=1}^{n} \frac{1}{F'(\alpha_r)} \int_{c_r}^{x} e^{\alpha_r(x-t)} f(t) \, dt \tag{45.5}$$

Example 45.1 To solve the equation of Example 44.3 by this method we notice that

$$\frac{1}{m^2+\omega^2} = \frac{1}{2\omega i}\left(\frac{1}{m-\omega i} - \frac{1}{m+\omega i}\right)$$

and hence that a particular integral is

$$(D^2+\omega^2)^{-1} \, e^{ax} \cos \omega x = \frac{1}{2\omega i} \int^{x} e^{i\omega(x-t)+at} \cos \omega t \, dt -$$

$$-\frac{1}{2\omega i} \int^{x} e^{-i\omega(x-t)+at} \cos \omega t \, dt =$$

$$= \frac{1}{\omega} \int^{x} \sin \{\omega(x-t)\} \cos \omega t \, e^{at} \, dt =$$

$$= \frac{1}{2\omega} \int^{x} e^{at} \, [\sin \omega x + \sin \{\omega(x-2t)\}] \, dt =$$

$$= \frac{1}{2\omega a} e^{ax} \sin \omega x + \frac{1}{2\omega(a^2+4\omega^2)} \times$$

$$\times e^{ax} \{2\omega \cos \omega x - a \sin \omega x\} =$$

$$= \frac{e^{ax}}{a^2+4\omega^2}\left\{\cos \omega x + \frac{2\omega}{a} \sin \omega x\right\}$$

as previously.

The relation of this method to that of the Laplace transform is obvious. Suppose that we wish to find the particular integral of $F(D)y = f(x)$ satisfying $y^{(r)}(0) = 0, r = 0, \ldots, n-1$. Then $\bar{y} = \mathscr{L} y$ satisfies $F(p)\bar{y}(p) =$

$\bar{f}(p)$, $\bar{f}=\mathscr{L}f$, so that

$$\bar{y}=\frac{\bar{f}(p)}{F(p)}=\sum_{r=1}^{n}\frac{1}{F'(\alpha_r)}\frac{\bar{f}(p)}{p-\alpha_r} \tag{45.6}$$

and hence

$$y(x)=\sum_{r=1}^{n}\frac{1}{F'(\alpha_r)}\mathscr{L}^{-1}[(p-\alpha_r)^{-1}\bar{f}(p);x]$$

Substituting $g(x)=e^{\alpha_r x}$, $\bar{g}(p)=(p-\alpha_r)^{-1}$ in (A1.16) we obtain the expression

$$y(x)=\sum_{r=1}^{n}\frac{1}{F'(\alpha_r)}\int_0^x e^{\alpha_r(x-t)}f(t)\,dt \tag{45.7}$$

for a particular integral of $F(D)y=f(x)$ when the auxiliary equation $F(m)=0$ has n distinct roots $\alpha_1,\alpha_2,\ldots,\alpha_n$.

If α is a root of order q of the equation $F(m)=0$, then the sum on the right-hand side of equation (45.6) will contain a term of the form

$$\frac{q!}{F^{(q)}(\alpha)}\cdot\frac{\bar{f}(p)}{(p-\alpha)^q}$$

Now from entry (2) of the table of Laplace transforms on p. 186 we deduce that

$$\mathscr{L}^{-1}[(p-\alpha)^{-q};x]=\frac{x^{q-1}}{(q-1)!}$$

and hence, from the convolution theorem, that this multiple zero of the auxiliary equation will contribute a term of the form

$$\frac{q}{F^{(q)}(\alpha)}\int_0^x (x-t)^{q-1}e^{\alpha(x-t)}f(t)\,dt$$

to a series of the type (45.7) for a particular solution of $F(D)y=f(x)$.

§46 Euler linear equations

An equation of the type

$$(x^nD^n+a_1x^{n-1}D^{n-1}+\cdots+a_{n-1}xD+a_n)y=f(x) \tag{46.1}$$

is called an *Euler linear equation*. It can be reduced to one with constant coefficients by changing the independent variable from x to t by the substitution $x = e^t$. By the chain rule

$$\frac{dy}{dx} = \frac{dy}{dt} \div \frac{dx}{dt} = \frac{1}{x}\frac{dy}{dt}.$$

If we write

$$D = \frac{d}{dx}, \qquad \theta = \frac{d}{dt}$$

we can rewrite this relation as

$$xD = \theta. \tag{46.2}$$

From this relation we deduce that

$$\theta^2 = xDxD = x^2D^2 + xD = x^2D^2 + \theta$$

and hence that

$$x^2D^2 = \theta(\theta - 1).$$

Continuing this process we find that if r is a positive integer

$$x^rD^r = \theta(\theta-1)\cdots(\theta-r+1).$$

Hence making this change of variable and writing $a_0 = 1$,

$$G(\theta) = \sum_{r=0}^{n} a_r\theta(\theta-1)\cdots(\theta-n+r+1) = \theta^n + \sum_{r=1}^{n} A_r\theta^{n-r},$$

$$g(t) = f(e^t)$$

we obtain the equation

$$G(\theta)y = g(t), \qquad \theta = \frac{d}{dt}$$

whose solution we have discussed in previous sections.

We observe that

$$(\theta-\alpha)(\theta-\beta)f = x^2f'' + x(1-\beta-\alpha)f' + \alpha\beta f = (\theta-\beta)(\theta-\alpha)f,$$

i.e. that the operators $\theta - \alpha$ and $\theta - \beta$ commute. Taking this result as basis, we may establish a calculus of the polynomial operator $G(\theta)$ and of its inverse $[G(\theta)]^{-1}$ similar to that of $F(D)$ and its inverse. In particular, since $\theta x^m = mx^m$ we deduce that

$$\theta^r x^m = m^r x^m$$

and hence that

$$G(\theta)x^m = G(m)x^m. \tag{46.3}$$

From this we deduce the inverse result

$$[G(\theta)]^{-1}x^m = x^m/G(m), \quad \text{if } G(m) \neq 0. \tag{46.4}$$

By direct differentiation we have

$$\theta x^m f(x) = mx^m f(x) + x^m \theta f(x) = x^m(m+\theta)f(x)$$

$$\theta^2 x^m f(x) = \theta\{x^m(m+\theta)f(x)\} = x^m(m+\theta)(m+\theta)f(x) =$$
$$= x^m(m+\theta)^2 f(x).$$

This suggests that if r is a positive integer

$$\theta^r x^m f(x) = x^m(m+\theta)^r f(x), \tag{46.5}$$

a result which is easily proved by induction. Hence, if $G(\theta)$ is a polynomial in θ, we have

$$G(\theta)x^m f(x) = x^m G(m+\theta)f(x). \tag{46.6}$$

Replacing $f(x)$ by $[G(m+\theta)]^{-1}f(x)$ we obtain the result

$$G(\theta)x^m[G(m+\theta)]^{-1}f(x) = x^m f(x),$$

which can be written in the alternative form

$$[G(\theta)]^{-1}x^m f(x) = x^m[G(m+\theta)]^{-1}f(x). \tag{46.7}$$

In particular

$$(\theta - m)^{-1}x^m = x^m\theta^{-1}(1),$$

where 1 denotes the unit function, and

$$(\theta - m)^{-r}x^m = x^m\theta^{-r}(1), \tag{46.8}$$

where $(\theta - m)^{-r}$ denotes the inverse of $(\theta - m)^r$. We therefore have to evaluate expressions of the form $\theta^{-r}(1)$.

Now

$$\theta \log|x| = x \cdot D\log|x| = x \cdot x^{-1} = 1$$

$$\theta^2\{\tfrac{1}{2}(\log|x|)^2\} = \theta \cdot \{x \log|x| x^{-1}\} = \theta \log|x| = 1$$

and, in general

$$\theta^r\left\{\frac{1}{r!}(\log|x|)^r\right\} = 1,$$

so that

$$\theta^{-r}(1) = \frac{1}{r!}(\log|x|)^r, \qquad r = 1, 2, \ldots. \tag{46.9}$$

Suppose now that

$$G(\theta) = G_r(\theta)(\theta - \alpha)^r,$$

where $G_r(\theta)$ is a polynomial of degree $n-r$ in θ, $G_r(a) \neq 0$; then

$$[G(\theta)]^{-1}x^\alpha = (\theta-\alpha)^{-r}[G_r(\theta)]^{-1}x^\alpha =$$

$$= (\theta-\alpha)^{-r}\frac{1}{G_r(\alpha)}x^\alpha =$$

$$= \frac{1}{G_r(\alpha)}x^\alpha\theta^{-r}(1) =$$

$$= \frac{1}{G_r(\alpha)}\frac{x^\alpha}{r!}(\log|x|)^r. \tag{46.10}$$

The complementary function of the differential equation

$$G(\theta)y = f(x), \tag{46.11}$$

$\theta = xD$, will contain terms like Cx^α, where α is a root of the indicial equation $G(m) = 0$. If this equation has n distinct roots, $m = \alpha, \beta, \ldots, \kappa$, the complementary function is

$$u(x) = c_1x^\alpha + c_2x^\beta + \cdots + c_nx^\kappa.$$

In the case of a root $m = \alpha$ repeated s times, the complementary function will contain a corresponding group of terms

$$x^\alpha\{c_1 + c_2\log|x| + \cdots + c_s(\log|x|)^{s-1}\}.$$

The most important type of inhomogeneous Euler linear equation (46.11) is that in which $f(x)$ is composed of terms like x^p and $x^p(\log|x|)^q$. A particular integral may be obtained by means of equations (46.4), (46.7) and (46.10), or by using a method of undetermined coefficients (see Example 46.2 below).

An equation of the more general type

$$(ax+b)^n y^{(n)} + a_1(ax+b)^{n-1}y^{(n-1)} + \cdots + a_{n-1}(ax+b)y' + a_n y = f(x)$$

may be brought into the above form by the linear substitution $ax + b = \xi$, or it may be transformed immediately into an equation with constant coefficients by changing the independent variable from x to t by the substitution $ax + b = e^t$.

Example 46.1 In terms of the operator θ the left-hand side of the equation

$$x^2y'' - 2xy' + 2y = x + x^2 \log|x| + x^3$$

may be written

$$G(\theta)y = \{\theta(\theta-1) - 2\theta + 2\}y = (\theta-1)(\theta-2)y.$$

The complementary function therefore is $c_1x + c_2x^2$.

To find a particular integral we notice that (46.10) yields

$$[G(\theta)]^{-1}x = -1 \cdot x \log|x|,$$

while (46.7) gives

$$[G(\theta)]^{-1}x^2 \log|x| = x^2\theta^{-1}(\theta+1)^{-1} \log|x|.$$

Since $\theta(\theta+1)\{\frac{1}{2}(\log|x|)^2\} = \log|x| + 1$, $\theta(\theta+1)(\log|x|) = 1$, we see that

$$(\theta+1)\theta\{\tfrac{1}{2}\log|x|^2 - \log|x|\} = \log|x|,$$

and hence that

$$\theta^{-1}(\theta+1)^{-1}\log|x| = \tfrac{1}{2}\log|x|^2 - \log|x|,$$

so

$$[G(\theta)]^{-1}x^2\log|x| = \tfrac{1}{2}x^2\log|x|^2 - x^2\log|x|.$$

Finally, from equation (46.4) with $G(3) = 2$, we see that

$$[G(\theta)]^{-1}x^3 = \tfrac{1}{2}x^3.$$

The general solution of the equation therefore is

$$y = c_1x + c_2x^2 - x\log|x| + x^2\{\tfrac{1}{2}(\log|x|)^2 - \log|x|\} + \tfrac{1}{2}x^3. \qquad (46.12)$$

If we change the independent variable from x to t where $x = \mathrm{e}^t$ the equation becomes

$$(\theta^2 - 3\theta + 2)y = \mathrm{e}^t + t\,\mathrm{e}^{2t} + \mathrm{e}^{3t},$$

where now $\theta = \mathrm{d}/\mathrm{d}t$. The complementary function is $c_1\,\mathrm{e}^t + c_2\,\mathrm{e}^{2t}$ and a particular integral is

$$(\theta-1)^{-1}(\theta-2)^{-1}(\mathrm{e}^t + t\,\mathrm{e}^{2t} + \mathrm{e}^{3t}) =$$

$$= -\mathrm{e}^t\,\theta^{-1}1 + \mathrm{e}^{2t}\,\theta^{-1}(\theta+1)^{-1}t + \tfrac{1}{2}\,\mathrm{e}^{3t} =$$

$$= -t\,\mathrm{e}^t + \mathrm{e}^{2t}\,(\theta^{-1} - 1)t + \tfrac{1}{2}\,\mathrm{e}^{3t} =$$

$$= -t\,\mathrm{e}^t + (\tfrac{1}{2}t^2 - t)\,\mathrm{e}^{2t} + \tfrac{1}{2}\,\mathrm{e}^{3t},$$

so the general solution is

$$y = c_1\,e^t + c_2\,e^{2t} - t\,e^t + (\tfrac{1}{2}t^2 - t)\,e^{2t} + \tfrac{1}{2}\,e^{3t}.$$

If now we write $e^t = x$, $t = \log|x|$, we get the solution (46.12).

Example 46.2 To solve the equation

$$(2x+1)^2 y'' + (4x+2)y' - 4y = x^2$$

we make the substitution $2x+1 = \xi$ to obtain

$$\xi^2 \frac{d^2 y}{d\xi^2} + \xi \frac{dy}{d\xi} - y = \tfrac{1}{16}(\xi - 1)^2,$$

which may be written

$$(\theta^2 - 1)y = \tfrac{1}{16}\xi^2 - \tfrac{1}{8}\xi + \tfrac{1}{16},$$

where now $\theta = \xi(d/d\xi)$. The complementary function is $c_1\xi + c_2/\xi$. Since ξ occurs in the complementary function we may take a particular integral of the form

$$z = a\xi^2 + b\xi \log|\xi| + c$$

for which

$$(\theta^2 - 1)z = 3a\xi^2 + 2b\xi - c,$$

giving $a = \frac{1}{48}$, $b = -\frac{1}{16}$, $c = -\frac{1}{16}$.

§47 Laplace linear equations

A linear equation of the type

$$(a_0 + b_0 x)y^{(n)} + (a_1 + b_1 x)y^{(n-1)} + \cdots + (a_n + b_n x)y = 0 \qquad (47.1)$$

in which the coefficients are linear functions of the independent variable is called a *Laplace linear equation*. If the constants $b_0, b_1, \ldots, b_n$ were all zero, the solution would be of exponential form. It was possibly this consideration that suggested the possibility of satisfying the equation by an integral of the form

$$I(x) = \int_\alpha^\beta f(t)\,e^{xt}\,dt \qquad (47.2)$$

in which f is a function of t alone and α, β are constants, as yet unspecified. In general, if D denotes d/dx,

$$D^r I = \int_\alpha^\beta f(t) t^r e^{xt} \, dt, \tag{47.3}$$

so that if the operator $\mathbf{L}$ is defined by

$$\mathbf{L} = (a_0 + b_0 x) D^n + (a_1 + b_1 x) D^{n-1} + \cdots + (a_n + b_n x),$$

we have

$$\mathbf{L} I(x) = \int_\alpha^\beta e^{xt} \{(a_0 + b_0 x) t^n + (a_1 + b_1 x) t^{n-1} + \cdots + a_n + b_n x\} f(t) \, dt$$

This equation can be written

$$\mathbf{L} I(x) = \int_\alpha^\beta e^{xt} \{P(t) + x Q(t)\} f(t) \, dt \tag{47.4}$$

where the functions P and Q are defined by the equations

$$P(t) = a_0 t^n + a_1 t^{n-1} + \cdots + a_n$$
$$Q(t) = b_0 t^n + b_1 t^{n-1} + \cdots + b_n.$$

Now

$$\int_\alpha^\beta x Q(t) e^{xt} f(t) \, dt = \int_\alpha^\beta \frac{\partial}{\partial t}(e^{xt}) Q(t) f(t) \, dt$$

so that, applying the rule for integrating by parts we see that the right side of this last equation can be written

$$[e^{xt} Q(t) f(t)]_\alpha^\beta - \int_\alpha^\beta e^{xt} \frac{d}{dt} \{Q(t) f(t)\} \, dt.$$

Using these results we see that equation (47.4) can be written

$$\mathbf{L} I(x) = [e^{xt} Q(t) f(t)]_\alpha^\beta - \int_\alpha^\beta [Q(t) f'(t) + \{Q'(t) - P(t)\} f(t)] e^{xt} \, dt.$$

From this equation we deduce that the function $I(x)$ defined by equation (47.2) is a solution of the Laplace linear equation (47.1) if

(i) $f(t)$ is a solution of the first-order equation

$$Q(t) f'(t) + \{Q'(t) - P(t)\} f(t) = 0;$$

(ii) $f(t)$ having been determined, the constants α and β are chosen in such a way that

$$[e^{xt}Q(t)f(t)]_\alpha^\beta = 0.$$

independently of x.

The differential equation in condition (i) has solution

$$f(t) = \frac{1}{Q(t)} \exp\left[\int \frac{P(t)}{Q(t)}\, dt\right], \tag{47.5}$$

so the condition (ii) becomes

$$[e^{xt} \exp\{\int (P/Q)\, dt\}]_\alpha^\beta = 0. \tag{47.6}$$

It should be observed that it is often advantageous to consider, instead of (47.2), a solution of the form

$$I_1(x) = \int_\alpha^\beta e^{ixt} f(t)\, dt \tag{47.7}$$

or one of the form

$$I_2(x) = \int_\alpha^\beta e^{-xt} f(t)\, dt. \tag{47.8}$$

In the case of (47.7) it is easily shown that

$$I_1(x) = -i[e^{ixt} Q_1(t) f(t)]_\alpha^\beta + \\ + i\int_\alpha^\beta [Q_1(t) f'(t) + \{Q_1'(t) - iP_1(t)\} f(t)]\, dt$$

where the functions $P_1(t)$ and $Q_1(t)$ are defined by the equations

$$P_1(t) = a_0(it)^n + a_1(it)^{n-1} + \cdots + a_n = P(it)$$

$$Q_1(t) = b_0(it)^n + b_1(it)^{n-1} + \cdots + b_n = Q(it)$$

Hence $I_1(x)$ is a solution of (47.1) if

(i)′ $f(t)$ is a solution of

$$Q_1(t) f'(t) + \{Q_1'(t) - iP_1(t)\} f(t) = 0.$$

(ii) α and β are such that

$$[e^{ixt} Q_1(t) f(t)]_\alpha^\beta = 0.$$

In a similar way we can show that $I_2(x)$ is a solution of (47.1) if

(i)″ $f(t)$ is a solution of

$$Q_2(t)f'(t)+\{Q'_2(t)+P_2(t)\}f(t)=0,$$

where $P_2(t)=P(-t)$, $Q_2(t)=Q(-t)$,

(ii)″ α and β are such that

$$[\mathrm{e}^{-xt}Q_2(t)f(t)]_\alpha^\beta=0.$$

Example 47.1 For the equation

$$xy''+(2n+1)y'+xy=0 \qquad (47.9)$$

we choose a solution of the form $I_1(x)$. In this case

$$P_1(t)=(2n+1)it, \qquad Q_1(t)=1-t^2$$

so that, applying the condition (i)′ we see that $f(t)$ must be chosen to be a solution of

$$(1-t^2)f'(t)+(2n-1)tf(t)=0,$$

i.e. we may take

$$f(t)=c(1-t^2)^{n-1/2}.$$

Since condition (ii)′ may be written

$$c[\mathrm{e}^{ixt}(1-t^2)^{n+1/2}]_\alpha^\beta=0$$

we see that we may take $\beta=1$, $\alpha=-1$. Therefore we have shown that

$$\int_{-1}^{1}(1-t^2)^{n-1/2}\,\mathrm{e}^{ixt}\,\mathrm{d}t$$

is a solution of the equation (47.9). Changing the variable of integration from t to θ where $t=\cos\theta$, we find that an equivalent solution is

$$c\int_0^\pi \cos(x\cos\theta)\sin^{2n}\theta\,\mathrm{d}\theta.$$

This integral is well known in the theory of Bessel functions. If we take $c=2^n n!/(2n)!\,\pi$ we find that the function

$$\frac{2^n n!}{(2n)!\pi}\int_0^\pi \cos(x\cos\theta)\sin^{2n}\theta\,\mathrm{d}\theta$$

is a solution of equation (47.9) and hence, by writing $y=x^{-n}w$ that

$$J_n(x)=\frac{2^n n!\, x^n}{\pi(2n)!}\int_0^\pi \cos(x\cos\theta)\sin^{2n}\theta\,\mathrm{d}\theta \tag{47.10}$$

is a solution of *Bessel's equation*

$$x^2w''+xw'+(x^2-n^2)w=0. \tag{47.11}$$

The function $J_n(x)$ is called *the Bessel function of the first kind of order n.*

If we replace $\cos(x\cos\theta)$ by its Maclaurin expansion and integrate term by term, making use of the formula

$$\int_0^\pi \cos^{2r}\theta\sin^{2n}\theta\,\mathrm{d}\theta=\left(\frac{1}{2}\right)^{2r+2n}\frac{(2r)!\,(2n)!}{r!\,n!\,(n+r)!}\pi,$$

we deduce from (47.10) that

$$J_n(x)=\sum_{r=0}^{\infty}\frac{(-1)^r(\frac{1}{2}x)^{2r+n}}{r!\,(n+r)!}. \tag{47.12}$$

It is easily shown that the series on the right-hand side of this equation is uniformly convergent for all values of x (see §55 below).

§48 Variation of parameters

When the complementary function of any linear equation

$$\mathsf{L}y=y^{(n)}+p_1y^{(n-1)}+\cdots+p_{n-1}y'+p_ny=f(x) \tag{48.1}$$

is known, the complete solution may be obtained by quadratures as follows. If the complementary function is

$$u(x)=\sum_{r=1}^{n}c_ru_r(x)$$

we replace the constants c_r by functions $v_r(x)$, which are as yet unspecified. Writing

$$y(x)=\sum_{r=1}^{n}v_r(x)u_r(x), \tag{48.2}$$

we proceed to set up n linear relations connecting the derivatives $v_1', v_2', \ldots, v_n'$. First of all we have

$$y'(x)=\sum_{r=1}^{n}v_r(x)u_r'(x)$$

provided that

$$\sum_{r=1}^{n} v_r'(x)u_r(x)=0.$$

Further

$$y''(x)=\sum_{r=1}^{n} v_r(x)u_r''(x)$$

provided that

$$\sum_{r=1}^{n} v_r'(x)u_r'(x)=0.$$

In general, we have

$$y^{(m)}(x)=\sum_{r=1}^{m} v_r(x)u_r^{(m)}(x), \quad 1<m<n-1,$$

provided that

$$\sum_{r=1}^{n} v_r'(x)u_r^{(m-1)}(x)=0, \qquad 1<m<n-1. \tag{48.3}$$

Finally,

$$y^{(n)}(x)=\sum_{r=1}^{n} v_r(x)u_r^{(n-1)}(x),$$

provided that

$$\sum_{r=1}^{n} v_r'(x)u_r^{(n-1)}(x)=f(x). \tag{48.4}$$

Hence, provided that $v_1', \ldots, v_n'$ satisfy the n equations (48.3) and (48.4)

$$\mathbf{L}y(x)=\sum_{m=0}^{n} p_m y^{(n-m)}, \qquad (p_0\equiv 1),$$

$$=\sum_{r=1}^{n} v_r \mathbf{L}u_r(x)=$$

$$=0$$

since $\mathbf{L}u_r(x)=0$, since u_r is a term in the complementary function. We now solve the n equations (48.3) and (48.4) for $v_1', v_2', \ldots, v_n'$ algebraically and obtain $v_1, v_2, \ldots, v_n$ by quadratures, thus obtaining a complete solution of the equation.

In the case of the second-order equation

$$y''+py'+qy=r, \tag{48.5}$$

we assume that u_1 and u_2 are the constituents of the complementary function and write for the general solution

$$y=v_1u_1+v_2u_2,$$

where

$$v_1'u_1+v_2'u_2=0, \qquad v_1'u_1'+v_2'u_2'=r.$$

The solution of these equations is

$$v_1'=-ru_2/W, \qquad v_2'=ru_1/W,$$

where $W(u_1u_2)=u_1u_2'-u_1'u_2$ is the Wronksian of u_1 and u_2 and is therefore non-zero. Hence the required general solution is

$$y(x)=-u_1\int\frac{ru_2}{W}\,dx+u_2\int\frac{ru_1}{W}\,dx. \tag{48.6}$$

Example 48.1 Since the complementary function of

$$y''+\omega^2y=f(x) \tag{48.7}$$

is $c_1\cos\omega x+c_2\sin\omega x$, we may take the solution to be

$$y(x)=v_1(x)\cos\omega x+v_2(x)\sin\omega x,$$

where

$$v_1'\cos\omega x+v_2'\sin\omega x=0, \qquad -\omega v_1'\sin\omega x+\omega v_2'\cos\omega x=f(x).$$

These equations have solution

$$v_1'(x)=-\omega^{-1}f(x)\sin\omega x, \qquad v_2'(x)=\omega^{-1}f(x)\cos\omega x.$$

Integrating the first of these two equations we have

$$v_1(x)=-\omega^{-1}\int_\alpha^x f(t)\sin\omega t\,dt$$

where α is an arbitrary constant. We can write this as

$$v_1(x)=c_1-\omega^{-1}\int_0^x f(t)\sin\omega t\,dt$$

where

$$c_1 = \omega^{-1} \int_0^{\alpha} f(t) \sin \omega t \, \mathrm{d}t$$

is an arbitrary constant.

Similarly, integrating the equation for v_2' we obtain

$$v_2(x) = c_2 + \omega^{-1} \int_0^x f(t) \cos \omega t \, \mathrm{d}t$$

where c_2 is an arbitrary constant. Hence the general solution of the differential equation (48.7) is

$$y(x) = c_1 \cos \omega x + c_2 \sin \omega x + \frac{1}{\omega} \int_0^x f(t) \sin \{\omega(x-t)\} \, \mathrm{d}t.$$

§49 Second-order linear equations: Green's functions

A second-order differential equation

$$\mathsf{L}y = 0$$

is said to be *exact* if there exists a first-order differential operator **M** such that

$$\mathsf{L}y = \frac{\mathrm{d}}{\mathrm{d}x}\{\mathbf{M}y\}.$$

In other words the second-order linear equation

$$p(x)y'' + q(x)y' + r(x)y = 0 \tag{49.1}$$

is exact if it can be written in the form

$$\frac{\mathrm{d}}{\mathrm{d}x}\{p(x)y' + s(x)y\} = 0. \tag{49.2}$$

Since

$$\frac{\mathrm{d}}{\mathrm{d}x}\{p(x)y'\} = p(x)y'' + p'(x)y'$$

$$\frac{\mathrm{d}}{\mathrm{d}x}\{q(x)y - p'(x)y\} = q(x)y' - p'(x)y' - \{p''(x) - q'(x)\}y,$$

we have the relation

$$\frac{d}{dx}\{p(x)y'+[q(x)-p'(x)]\,y\}=$$

$$=p(x)y''+q(x)y'+r(x)y-\{p''(x)-q'(x)+r(x)\}y,$$

so that the equation (49.1) is exact if and only if

$$p''(x)-q'(x)+r(x)=0 \tag{49.3}$$

for all x, and can be written then in the form (49.2) with

$$s(x)=q(x)-p'(x). \tag{49.4}$$

We solve an exact equation of this type by integrating (49.2) to obtain the first-order linear equation

$$p(x)y'+s(x)y=c_1, \tag{49.5}$$

where c_1 is an arbitrary constant. The solution of this equation, in turn, involves a second arbitrary constant c_2. Equation (49.3) is called a first integral of the equation (49.1).

If the equation $\mathbf{L}y=0$ is not exact, but if there exists a function $z(x)$ such that $z\mathbf{L}y=0$ is exact, we say that $z(x)$ is an *integrating factor* of the differential equation $\mathbf{L}y=0$. For the differential operator

$$\mathbf{L}=p(x)\frac{d^2}{dx^2}+q(x)\frac{d}{dx}+r(x), \tag{49.6}$$

we see from equation (49.3) that $z(x)$ is an integrating factor of the equation $\mathbf{L}y=0$ if it is a solution of the differential equation

$$\mathbf{L}^*z=0, \tag{49.7}$$

where the operator $\mathbf{L}^*$ is defined by the equation

$$\mathbf{L}^*z=\frac{d}{dx^2}\{p(x)z\}-\frac{d}{dx}\{q(x)z\}+r(x)z. \tag{49.8}$$

The operator $\mathbf{L}^*$ is called the *adjoint* of the operator $\mathbf{L}$.

If we carry out the differentiations we find that

$$\mathbf{L}^*z=p(x)z''+\{2p'(x)-q(x)\}z'+\{p''(x)-q'(x)+r(x)\}z,$$

so that

$$(\mathbf{L}^*)^* = \frac{d^2}{dx^2}(pz) - \frac{d}{dx}(2p'-q)z + (p''-q'+r)z =$$

$$= (p''z + 2p'z' + pz'') - (2p''-q')z - (2p'-q)z' + (p''-q'+r)z =$$

$$= p''z + qz' + rz;$$

in other words, we have shown that

$$(\mathbf{L}^*)^* = \mathbf{L}. \tag{49.9}$$

It follows immediately from the definition that if $\mathbf{L}_1$ and $\mathbf{L}_2$ are second-order linear differential operators

$$(\mathbf{L}_1 + \mathbf{L}_2)^* = \mathbf{L}_1^* + \mathbf{L}_2^*$$

$$(k\mathbf{L}_1)^* = k\mathbf{L}_1^*$$

where k is any real number.

If a differential operator $\mathbf{L}$ has the property that $\mathbf{L}^* = \mathbf{L}$, it is said to be *self-adjoint* or *formally self-adjoint.*

We suppose now that $p \in C^2[a,b]$, $q \in C^1[a,b]$, $r \in C[a,b]$. Then if y and z are arbitrary functions in $C^2[a,b]$ we find, on integrating by parts, that

$$\int_a^b qzy'\,dx = q_b z_b y_b - q_a z_a y_a - \int_a^b y\frac{d}{dx}(qz)\,dx,$$

where we have introduced the notation

$$y_a = y(a), \qquad y'_a = y'(a), \qquad y_b = y(b), \qquad y'_b = y'(b),$$

and that

$$\int_a^b pzy''\,dx = p_b z_b y'_b - p_a z_a y'_a - (p_b z'_b + p'_b z_b)\,y_b + (p_a z'_a + p'_a z_a)y_a +$$

$$+ \int_a^b y\frac{d^2}{dx^2}(pz)\,dx.$$

If we introduce the inner product

$$\langle f, g\rangle = \int_a^b f(x)g(x)\,dx$$

we deduce from these expressions that

$$\langle \mathbf{L}y, z\rangle - \langle y, \mathbf{L}^* z\rangle = P(\eta, \zeta) \tag{49.10}$$

where $P(\zeta, \eta)$ is the bilinear form in the variables $\eta = (y_a, y'_a, y_b, y'_b)$, $\zeta = (z_a, z'_a, z_b, z'_b)$ defined by the equation

$$P(\eta, \zeta) = p_b z_b y'_b - p_a z_a y'_a - (p_b z'_b + p'_b z_b) y_b + (p_a z'_a + p'_a z_a) y_a + \\ + q_b z_b y_b - q_a z_a y_a. \tag{49.11}$$

The equation (49.10) is known as *Lagrange's identity*.

We shall now consider the initial-value problem of the equation

$$\mathbf{L} y(x) = f(x), \qquad x \geqslant a, \tag{49.12}$$

where $\mathbf{L}$ denotes the linear operator

$$\mathbf{L} = \frac{\mathrm{d}^2}{\mathrm{d}x^2} + p(x) \frac{\mathrm{d}}{\mathrm{d}x} + q(x).$$

The initial-value problem is that of finding the twice-differentiable function which satisfies the differential equation (49.12) and the initial conditions

$$y(a) = y_0, \qquad y'(a) = y_1, \tag{49.13}$$

where y_0 and y_1 are real numbers. The problem may be split into two parts:

(i) Find a solution of the homogeneous equation $\mathbf{L} y(x) = 0$ satisfying the initial conditions (49.13);

(ii) Find a solution $u(x)$ of the inhomogeneous equation (49.12) satisfying the initial conditions

$$u(a) = u'(a) = 0. \tag{49.14}$$

By the *Green's function* $G(x, t)$ of the operator $\mathbf{L}$ for the initial-value problem at $x = a$ is meant a function such that

$$u(x) = \int_a^x G(x, t) f(t) \, \mathrm{d}t \qquad (x \geqslant a), \tag{49.15}$$

is the solution of the initial-value problem posed by the differential equation (49.12) and the initial conditions (49.14).

For a given operator $\mathbf{L}$ of the above form we can determine the Green's function by means of the following theorem, in the statement of which we use the following notation:

$$\mathbf{L}_x = \frac{\partial^2}{\partial x^2} + p(x) \frac{\partial}{\partial x} + q(x).$$

Theorem 49.1 *Suppose that $G(x, t)$ is, for fixed $t \geqslant a$, the solution of*

$$\mathsf{L}_x G(x, t) = 0 \qquad (x \geqslant a),$$

satisfying the conditions

$$G(t, t) = 0, \qquad G_x(t, t) = 1 \qquad (t \geqslant a).$$

Then $G(x, t)$ is the Green's function of the operator L for the initial-value problem at $x = a$.

The proof is direct. We consider

$$u(x) = \int_a^x G(x, t) f(t)\, \mathrm{d}t.$$

Then

$$u'(x) = \int_a^x G_x(x, t) f(t)\, \mathrm{d}t + G(x, x) f(x),$$

and since $G(x, x) = 0$, this reduces to

$$u'(x) = \int_a^x G_x(x, t) f(t)\, \mathrm{d}t.$$

Differentiating again we obtain

$$u''(x) = \int_a^x G_{xx}(x, t) f(t)\, \mathrm{d}t + G_x(x, x) f(x).$$

Since $G_x(x, x) = 1$ this reduces to

$$u''(x) = \int_a^x G_{xx}(x, t) f(t)\, \mathrm{d}t + f(x).$$

Since the operator L is linear in u, u' and u'' we deduce immediately that equation (49.12) is satisfied. It is immediately obvious from the above integral expressions for $u(x)$ and $u'(x)$ that $u(a) = u'(a) = 0$.

It is now a simple matter to construct this Green's function in cases in which two linearly independent solutions of the homogeneous equation $\mathsf{L}y = 0$ are known; if we denote them by $y_1(x)$ and $y_2(x)$, then we may write

$$G(x, t) = c_1(t) y_1(x) + c_2(t) y_2(x)$$

where

$$0 = c_1(t) y_1(t) + c_2(t) y_2(t)$$
$$1 = c_1(t) y_1'(t) + c_2(t) y_2'(t)$$

from which we deduce the formula

$$G(x,t)=\frac{y_1(x)y_2(t)-y_2(x)y_1(t)}{y_1'(t)y_2(t)-y_1(t)y_2'(t)}. \tag{49.16}$$

Example 49.1 Find the Green's function for the initial-value problem for the operator

$$\frac{\mathrm{d}^2}{\mathrm{d}x^2}+\omega^2.$$

In this case we may take $y_1(x)=\sin\omega x$, $y_2(x)=\cos\omega x$, so that

$$G(x,t)=\frac{\sin\omega x\cos\omega t-\sin\omega t\cos\omega x}{\omega(\cos^2\omega t+\sin^2\omega t)},$$

an equation which may be written in the alternative form

$$G(x,t)=\frac{\sin\omega(x-t)}{\omega}.$$

From this it follows that the solution of the initial-value problem

$$y''+\omega^2 y=0 \quad (x\geqslant 0), \qquad y(0)=a, \quad y'(0)=U$$

is

$$y(x)=a\cos\omega x+(U/\omega)\sin\omega x+\frac{1}{\omega}\int_0^x f(t)\sin\{\omega(x-t)\}\,\mathrm{d}t.$$

The problem of determining a function $y(x)$ such that

$$\mathbf{L}y(x)=f(x), \qquad a\leqslant x\leqslant b, \tag{49.17}$$

$$\lambda_1 y(a)+\mu_1 y'(a)=\nu_1, \qquad \lambda_2 y(b)+\mu_2 y'(b)=\nu_2 \tag{49.18}$$

is called a *two-point boundary-value problem.*

This problem, too, can be split into two parts:

(i) Find a solution of $\mathbf{L}y=0$ satisfying the boundary conditions (49.18);
(ii) Find a solution of $\mathbf{L}u=f$ satisfying the boundary conditions

$$\lambda_1 y(a)+\mu_1 y'(a)=0, \qquad \lambda_2 y(b)+\mu_2 y'(b)=0. \tag{49.19}$$

The function $G(x,t)$ with the property that the solution of this two-point boundary-value problem can be written in the form

$$u(x)=\int_a^b G(x,t)f(t)\,\mathrm{d}t$$

is called the Green's function for the two-point boundary value problem posed by equations (49.17) and (49.18). The procedure for calculating the Green's function in this second case is contained in:

Theorem 49.2 *Suppose that the function $G(x,t)$ satisfies the differential equation $\mathbf{L}_x G(x,t)=0$, the boundary conditions*

$$\lambda_1 G(a,t)+\mu_1 G_x(a,t)=0, \qquad \lambda_2 G(b,t)+\mu_2 G_x(b,t)=0 \tag{49.20}$$

and the conditions

$$G(t+,t)=G(t-,t), \qquad G_x(t+,t)-G_x(t-,t)=1. \tag{49.21}$$

Then $G(x,t)$ is the Green's function for the two-point boundary-value problem posed by equations (49.17) *and* (49.18).

The proof is again simple. Since $G(x,t)$ is not continuously differentiable, we write

$$u(x)=\int_a^x G(x,t)f(t)\,\mathrm{d}t+\int_x^b G(x,t)f(t)\,\mathrm{d}t.$$

Then, differentiating, we obtain

$$u'(x)=\int_a^x G_x(x,t)f(t)\,\mathrm{d}t+G(x,x-)f(x)+$$

$$+\int_x^b G_x(x,t)f(t)\,\mathrm{d}t-G(x,x+)f(x).$$

From the first of equations (49.21) we see that this reduces to

$$u'(x)=\int_a^x G_x(x,t)f(t)\,\mathrm{d}t+\int_x^b G_x(x,t)f(t)\,\mathrm{d}t.$$

Repeating the argument, we find that

$$u''(x)=\int_x^x G_{xx}(x,t)f(t)\,\mathrm{d}t+\int_x^b G_{xx}(x,t)f(t)\,\mathrm{d}t+$$

$$+\{G_x(x,x-)-G_x(x,x+)\}f(x)$$

$$u''(x)=\int_a^b G_{xx}(x,t)f(t)\,\mathrm{d}t+f(x),$$

so that

$$\mathbf{L}u(x)=f(x)+\int \mathbf{L}_x G(x,t)f(t)\,\mathrm{d}t.$$

The condition $\mathbf{L}_x G(x,t)=0$ implies that $\mathbf{L}u=f$ and

$$\lambda_1 u(a)+\mu_1 u'(a)=\int_a^b \{\lambda_1 G(a,t)+\mu_1 G_x(a,t)\}f(t)\,\mathrm{d}t$$

by the first of equations (49.20); the second condition is similarly satisfied.

Suppose now that $y_1(x)$ and $y_2(x)$ are two linearly independent solutions of $\mathbf{L}y=0$ satisfying the conditions

$$\lambda_1 y_1(a)+\mu_1 y_1'(a)=0, \qquad \lambda_2 y_2(b)+\mu_2 y_1'(b)=0,$$

and let

$$G(x,t)=\begin{cases} c_1(t)y_2(x), & a\leqslant t\leqslant x \\ c_2(y)y_1(x), & x\leqslant t\leqslant b \end{cases}$$

where

$$c_1(t)y_2(t)=c_2(t)y_1(t), \qquad -c_2(t)y_1'(t)+c_1(t)y_2'(t)=1.$$

Solving these equations for c_1 and c_2 we therefore obtain the formulae

$$G(x,t)=\begin{cases} \dfrac{y_1(t)y_2(x)}{W(y_1,y_2;t)}, & a\leqslant t\leqslant x; \\ \dfrac{y_2(t)y_1(x)}{W(y_1,y_2;t)}, & x\leqslant t\leqslant b. \end{cases}$$

In other words, we have

$$u(x)=y_2(x)\int_a^x \frac{y_1(t)f(t)}{W(y_1,y_2;t)}\,\mathrm{d}t+y_1(x)\int_x^b \frac{y_2(t)f(t)}{W(y_1,y_2;t)}\,\mathrm{d}t. \quad (49.22)$$

We illustrate the method by considering:

Example 49.2 Find the function $y(x)$ satisfying

$$(1-x^2)y''(x)+2xy'(x)-2y(x)=f(x), \qquad -1\leqslant x\leqslant 1$$

$$y(-1)=y(1)=0.$$

In this case

$$\mathbf{L}y=y''(x)+\frac{2x}{1-x^2}y(x)-\frac{2}{1-x^2}y(x).$$

It is easily seen that we may take $y_1(x)=(1+x)^2$, $y_2(x)=(1-x)^2$, for which $W(y_1,y_2;t)=-4(1-t^2)$; since the given differential equation may be written

$$\mathbf{L}y(x) = (1-x^2)^{-1} f(x),$$

we deduce from equation (49.22) that the required solution is

$$y(x) = -\tfrac{1}{4}(1-x)^2 \int_{-1}^{x} \frac{f(t)\,\mathrm{d}t}{(1-t)^2} - \tfrac{1}{4}(1+x)^2 \int_{x}^{1} \frac{f(t)\,\mathrm{d}t}{(1+t)^2}.$$

6

Solution in series

§50 Solution developed as a Taylor series

In this section we consider a linear equation of the second order

$$y'' = p(x)y' + q(x)y \tag{50.1}$$

in which we shall regard x as a complex variable.

This type includes many equations of very great importance which cannot be solved in terms of simple combinations of elementary functions. Given such an equation, the usual procedure is to express the solution (which may be the solution satisfying certain initial conditions) in the form of an infinite series from which tables of the value of the solution may, if desired, be computed. Thus the convergence of the series is important, not only as a basis of the validity of the process, but also as an indication of the practical value of the result, for a slowly converging series is of little value for purposes of calculation.

It will be assumed that the coefficients $p(x)$, $q(x)$ are single valued, and have derivatives of all orders except possibly for certain isolated values of x. We shall suppose that $x=a$ is a value for which p, q and all their derivatives are finite: we shall obtain a solution such that $y(a)=y_0$, $y'(a)=y_1$, where y_0 and y_1 are assigned finite values.

By substituting these values in equation (50.1) we obtain the corresponding value of the second derivative, namely

$$y''(a) = p(a)y_1 + q(a)y_0.$$

Differentiating (50.1) we obtain

$$y''(a) = p(a)y''(a) + \{p'(a) + q(a)\}y_1 + q'(a)y_0$$

and as y_0, y_1 and $y''(a)$ are known, $y''(a)$ can be obtained immediately. Continuing the process, we obtain the values of successive derivatives for $x=a$, and thus we have the coefficients in the Taylor series

$$y(x) = y_0 + y_1(x-a) + y''(a)\frac{(x-a)^2}{2!} + \cdots + y^{(n)}(a)\frac{(x-a)^n}{n!} + \cdots.$$

Borrowing from the language of the theory of functions of a complex variable, a point $x=a$ at which the above conditions are satisfied is said to be an *ordinary point* of the equation.

Example 50.1 To find the solution of the equation

$$y''+xy=0,$$

such that $y=c_1, y'=c_2$ when $x=0$.

We have

$$y''=-xy$$
$$y'''=-y-xy'$$

showing that $y''(0)=0$, $y'''(0)=-c_1$. Differentiating both sides of the first of these equations $(n-2)$ times with respect to x and using Leibnitz's theorem we have

$$y^{(n)}(x)=xy^{(n-2)}(x)-(n-2)y^{(n-3)} \qquad (n\geqslant 2).$$

The origin is an ordinary point; putting $x=0$ in the above equations, we have

$$y''(0)=0, \qquad y'''(0)=-c_1, \qquad y^{iv}(0)=-2c_2, \qquad y^{v}(0)=0,$$

and in general

$$y^{(n)}(0)=-(n-2)y^{(n-3)}(0).$$

We thus find that at $x=0$,

$$y^{(3n)}=(-1)(-4)(-7)\cdots(-3n+2)c_1,$$
$$y^{(3n+1)}=(-2)(-5)(-8)\cdots(-3n+1)c_2,$$
$$y^{(3n+2)}=0.$$

Thus the required solution is

$$y(x)=c_1y_1(x)+c_2y_2(x),$$

where the functions $y_1(x)$ and $y_2(x)$ are defined by the infinite series

$$y_1(x)=1-\frac{x^3}{3!}+\frac{4x^6}{6!}-\frac{28x^9}{9!}+\cdots$$

$$y_2(x)=x-\frac{2x^4}{4!}+\frac{10x^7}{7!}-\frac{80x^{10}}{10!}+\cdots$$

and it may be easily verified that (i) the ratio test proves the convergence of each series for all values of x, (ii) each series separately satisfies the differential equation. Since c_1 and c_2 may be regarded as arbitrary constants, the above is an expression of the general solution.

Note 1. When $p(x)$, $q(x)$ are polynomials of low degree, a more practical way of obtaining the solution is to assume a series of ascending powers of x

$$y(x)=\sum_{r=0}^{\infty} c_r x^r, \tag{50.2}$$

to substitute this series in the differential equation and to determine the coefficients so that the equation is identically satisfied; that is to say, so that the coefficient of each individual power of x cancels out.

Example 50.2 Substitution from (50.2) into the equation

$$y''-xy'+ny=0$$

and equating to zero the coefficients of successive powers of x in the result, we obtain the following set of recurrence relations:

$$2c_2+nc_0=0,$$

$$6c_3+(n-1)c_1=0,$$

$$12c_4+(n-2)c_2=0,$$

$$20c_5+(n-3)c_3=0,$$

and, in general, by equating to zero the coefficient of x^r,

$$(r+1)(r+2)c_{r+2}+(n-r)c_r=0 \qquad (r=0, 1, 2, \ldots).$$

We see that the coefficients of even rank (i.e. suffix) depend on c_0 but not on c_1; those of odd rank on c_1 but not on c_0, and that c_0 and c_1 are arbitrary. Denoting by $S_1(x)$ the series obtained by taking $c_0=1, c_1=0$ and by $S_2(x)$ the series obtained by taking $c_0=0$, $c_1=1$, we have

$$S_1(x)=1-\frac{n}{2!}x^2+\frac{n(n-2)}{4!}x^4-\frac{n(n-2)(n-4)}{6!}x^6+\cdots$$

$$S_2(x)=x-\frac{n-1}{3!}x^3+\frac{(n-1)(n-3)}{5!}x^5-\frac{(n-1)(n-3)(n-5)}{7!}x^7+\cdots.$$

Since

$$\frac{c_{r+2}}{c_r}=\frac{r-n}{(r+1)(r+2)}\to 0 \qquad \text{as } r\to\infty$$

for any finite value of n, both series converge for all values of x. The general solution of the equation is

$$y(x) = AS_1(x) + BS_2(x),$$

where A, B are arbitrary constants. It will be noted that when n is an even positive integer the series S_1 terminates and reduces to a polynomial of degree n; when n is odd, S_2 is a polynomial of degree n.

Note 2. It is sometimes possible, by a change of dependent variable, to simplify a linear equation, or to identify it with one previously studied. The transformation most frequently employed is of the form $y = uv$, where y is the original, v the new dependent variable, and u a chosen function of x.

Example 50.3 If the equation

$$y'' + (n + \tfrac{1}{2} - \tfrac{1}{4}x^2)y = 0$$

is satisfied by a series of ascending integral powers of x of the type (50.2), the recurrence relation between the coefficients is

$$(r+1)(r+2)c_{r+2} + (n + \tfrac{1}{2})c_r - \tfrac{1}{4}c_{r-2} = 0.$$

This is a three-term recurrence relation which is much more difficult to manipulate than one that involves only two terms. If, however, we make the substitution

$$y(x) = e^{-\frac{1}{4}x^2}\, v(x),$$

we find that the equation in v is

$$v'' - xv' + nv = 0,$$

which is precisely the equation considered above.

In the above examples we derived solutions in the form of Maclaurin series since $x = 0$ was, in each case, an ordinary point of the differential equation under consideration. As an example of an equation which has a solution in the form of a Taylor series we consider:

Example 50.4 To find the solution of *Legendre's equation*

$$(1 - x^2)y'' - 2xy' + v(v+1)y = 0 \tag{50.3}$$

satisfying the initial condition $y(1) = 1$, we differentiate both sides of the equation $(r-1)$ times with respect to x, making use of Leibnitz's theorem to obtain

$$(1-x^2)y^{(r+1)}(x)-2rxy^{(r)}(x)+(v-r)(v+1+r)y^{(r-1)}(x)=0.$$

Putting $x=1$ we obtain the recurrence relation

$$y^{(r)}(1)=\frac{(v+r)(v+1-r)}{2r}y^{(r-1)}(1)$$

from which we derive the formula

$$y^{(r)}(1)=\frac{(v+1)(v+2)\cdots(v+r)(-v)(-v+1)\cdots(-v+r-1)}{r!}(-\tfrac{1}{2})^r.$$

Introducing the Pochhammer symbol $(a)_r$ defined by the equations

$$(a)_r=a(a+1)(a+2)\cdots(a+r-1), \qquad (a)_0=1$$

we can rewrite this formula as

$$y^{(r)}(1)=\frac{(-v)_r(v+1)_r}{(1)_r}(-\tfrac{1}{2})^r,$$

since $(1)_r=r!$ Substituting this expression in the Taylor expansion

$$y(x)=\sum_{r=0}^{\infty}y^{(r)}(1)\frac{(x-1)^r}{r!}$$

we obtain the solution

$$y(x)=\sum_{r=0}^{\infty}\frac{(-v)_r(v+1)_r}{(1)_r r!}\left(\frac{1-x}{2}\right)^r, \qquad (50.4)$$

which is convergent if $|\tfrac{1}{2}-\tfrac{1}{2}x|<1$, i.e. $-\tfrac{1}{2}<x<\tfrac{3}{2}$.

This is a convenient point at which to introduce the notation for a hypergeometric series. We write

$${}_2F_1(a,b;c;x)=\sum_{r=0}^{\infty}\frac{(a)_r(b)_r}{(c)_r r!}x^r. \qquad (50.5)$$

With this notation we can re-write equation (50.4) as

$$y(x)={}_2F_1(-v,v+1;1;\tfrac{1}{2}-\tfrac{1}{2}x). \qquad (50.6)$$

It should be noticed that if v is a positive integer, n say, $(-n)_{n+1}=0$ showing that the polynomial

$$P_n(x)={}_2F_1(-n,n+1;1;\tfrac{1}{2}-\tfrac{1}{2}x) \qquad (50.7)$$

is a solution of Legendre's equation of order n. It is called the *Legendre polynomial of degree n.*

To obtain a second solution of Legendre's equation we use a device which is applicable to self-adjoint equations. If $y_1(x)$ is a solution of

$$\frac{\mathrm{d}}{\mathrm{d}x}\left(p(x)\frac{\mathrm{d}y}{\mathrm{d}x}\right)+q(x)y(x)=0, \tag{50.8}$$

we substitute $y=y_1 v$ to obtain the differential equation

$$py_1 v''+(2py_1'+p'y_1)v'=0.$$

Noting that this equation has an integrating factor y_1, we can write it as

$$\frac{\mathrm{d}}{\mathrm{d}x}(py^2v')=0$$

with solution

$$v=c_1+c_2\int\frac{\mathrm{d}x}{p(x)\{y_1(x)\}^2}.$$

Hence the general solution of (50.8) can be written

$$y(x)=c_1 y_1(x)+c_2 y_2(x),$$

with

$$y_2(x)=y_1(x)\int\frac{\mathrm{d}x}{p(x)\{y_1(x)\}^2}. \tag{50.9}$$

We therefore find by taking $y_1(x)=P_n(x)$, $p(x)=1-x^2$, that the general solution of Legendre's equation is

$$y(x)=c_1P_n(x)+c_2Q_n(x), \tag{50.10}$$

where $P_n(x)$ is defined by equation (50.7) and $Q_n(x)$ is defined by the equation

$$Q_n(x)=P_n(x)\int\frac{\mathrm{d}x}{(1-x^2)\{P_n(x)\}^2}. \tag{50.11}$$

From equation (50.7) we see that

$$P_n(x)=1-\tfrac{1}{2}n(n+1)(1-x)+O\{(1-x)^2\}$$

and hence that

$$\frac{1}{\{P_n(x)\}^2}=1+n(n+1)(1-x)+O\{(1-x)^2\}$$

and

$$\int \frac{dx}{(1-x^2)\{P_n(x)\}^2} = \int \left[\frac{1}{1-x^2} + \frac{n(n+1)}{1+x}\right] dx =$$

$$= \log\left|\frac{1+x}{1-x}\right| + n(n+1)\log|x+1| \qquad \to \infty \quad \text{as } x \to 1.$$

Hence we deduce that the only solutions of Legendre's equation which remain finite as $x \to 1$ are of the form

$$y = cP_n(x); \tag{50.12}$$

the same result is true for solutions remaining finite as $x \to -1$.

§51 Regular singular points

When we consider an equation of the form

$$P(x)y'' + Q(x)y' + R(x)y = 0 \tag{51.1}$$

we see that if $x-a$ is a factor of the coefficient $P(x)$, but not of both $Q(x)$ and $R(x)$, the equation fails to determine a value of y'' for $x=a$ and the procedure of the last section is inapplicable. We then say that the point $x=a$ is a *singular point* or a *singularity* of the equation.

The form of a solution valid in the neighbourhood of a singularity may therefore differ from that of a solution appropriate to an ordinary point. It is important to investigate this difference, which is not merely a matter of theoretical interest, but of practical importance, as the solutions valid in the neighbourhood of singular points are as a general rule those required in physical applications.

We find a clue to a fruitful line of approach in the equation

$$(x-a)^2 y'' + b(x-a)y' + cy = 0,$$

where a, b and c are real constants. This equation has a solution of the form

$$y = (x-a)^\sigma,$$

where σ is a root of the equation

$$\sigma^2 + (b-1)\sigma + c = 0,$$

so that the singularity of the equation may imply a singularity (e.g. a pole or a branch point) in its solution. When equation (51.1) can be written in the form

$$(x-a)^2 p(x)y'' + (x-a)q(x)y' + r(x)y = 0,$$

with $p(a) \neq 0$, and $p(x)$, $q(x)$ and $r(x)$ are finite in the neighbourhood of $x = a$, it is thus natural to assume a solution of the form

$$y = (x-a)^{\sigma} F(x),$$

where $F(x)$ is a function which can be developed as a Taylor series in $(x-a)$; in other words it is natural to assume a series solution of the type

$$y = \sum_{r=1} c_r (x-a)^{\sigma+r}.$$

It will be found on substitution and then equating to zero the lowest power of $x-a$, that σ is determined by the *indicial equation* of the second degree

$$\sigma(\sigma-1)p(a) + \sigma q(a) + r(a) = 0,$$

giving in general two distinct values of σ and leading to two distinct solutions. Such a singularity is called a *regular singular point*.

The procedure is straightforward as long as the roots of the indicial equation are distinct and do not differ by an integer so we shall consider that case first, and then look at the case of equal roots.

§52 The hypergeometric equation

The equation

$$x(1-x)y'' + \{\gamma - (\alpha + \beta + 1)x\}y' - \alpha\beta y = 0, \tag{52.1}$$

in which α, β and γ are constants, is called the hypergeometric equation. On multiplying throughout by x, it will be seen that $x = 0$ is a regular singular point. Similarly, the point $x = 1$ is a regular singular point and there are no others.

Confining our attention to the singular point $x = 0$, we assume the series

$$y(x) = \sum_{r=0}^{\infty} c_r x^{\sigma+r}, \tag{52.2}$$

in which the index σ is, for the moment, regarded as an arbitrary parameter. Denoting the left-hand member of the equation (52.1) by $\mathbf{L}y$, we find on substituting the series (52.2) for y that

$$\mathsf{L}y=\sigma(\sigma-1+\gamma)c_0x^{\sigma-1}+\sum_{r=0}^{\infty}\{(\sigma+r+1)(\sigma+r+\gamma)c_{r+1}-$$
$$-(\sigma+r+\alpha)(\sigma+r+\beta)c_r x^{\sigma+r}.$$

If now, whatever σ may be, we choose the coefficients $c_1, c_2, \ldots$ so that

$$(\sigma+r+1)(\sigma+r+\gamma)c_{r+1}=(\sigma+r+\alpha)(\sigma+r+\beta)c_r \qquad (r\geqslant 0), \quad (52.3)$$

leaving c_0 arbitrary, we find that

$$c_1=\frac{(\sigma+\alpha)(\sigma+\beta)}{(\sigma+1)(\sigma+\gamma)}c_0, \qquad c_2=\frac{(\sigma+\alpha)(\sigma+\alpha+1)(\sigma+\beta)(\sigma+\beta+1)}{(\sigma+1)(\sigma+2)(\sigma+\gamma)(\sigma+\gamma+1)}c_0,$$

and so on, and these coefficients are finite provided that neither σ nor $\sigma+\gamma-1$ is a negative integer. Setting such cases aside we have proved that if

$$\phi(x,\sigma)=c_0x^{\sigma}\left\{1+\sum_{r=1}^{\infty}\frac{(\sigma+\alpha)_r(\sigma+\beta)_r}{(\sigma+1)_r(\sigma+\gamma)_r}x^r\right\}, \qquad (52.4)$$

and if the series is convergent, then

$$\mathsf{L}\phi(x,\sigma)=\sigma(\sigma-1+\gamma)c_0x^{\sigma-1}.$$

The question of convergence can be settled once and for all by the ratio test, for

$$\frac{c_{r+1}}{c_r}=\frac{(\sigma+r+\alpha)(\sigma+r+\beta)}{(\sigma+r+1)(\sigma+r+\gamma)}\to 1 \qquad \text{as } r\to\infty$$

$\sigma, \alpha, \beta, \gamma$ being finite, and so the series converges for $|x|<1$.

We can make $\mathsf{L}\phi(x,\sigma)=0$, and thus satisfy the equation, by taking

$$\sigma(\sigma-1+\gamma)=0 \qquad (52.5)$$

provided that the stipulation that neither σ nor $\sigma+\gamma-1$ is a negative integer is not violated. Thus the roots of the indicial equation are $\sigma=0$ provided $\gamma-1$ is not a negative integer, and $1-\gamma$ provided $1-\gamma$ is not a negative integer.

Consider first the index $\sigma=0$. A solution is provided by

$$\phi(x,0)=c_0\cdot {}_2F_1(\alpha,\beta;\gamma;x)$$

provided that γ is not zero or a negative integer. Next take the alternative index $\sigma=1-\gamma$, which gives the solution

$$\phi(x,1-\gamma)=c_0x^{1-\gamma}{}_2F_1(\alpha-\gamma+1,\beta-\gamma+1;2-\gamma;x)$$

provided γ is not a positive integer greater than unity. We shall suppose that γ is not an integer, positive or negative or zero; then the general solution of the hypergeometric equation (52.1) is

$$y = A\,{}_2F_1(\alpha, \beta, \gamma; x) + Bx^{1-\gamma}\,{}_2F_1(\alpha-\gamma+1, \beta-\gamma+1; 2-\gamma; x), \quad (52.6)$$

where A and B are arbitrary constants.

The above procedure is known as *the Method of Frobenius.*

If we change the independent variable in the hypergeometric equation (52.1) from x to $t = 1-2x$ we find that the equation becomes

$$t(1-t)\frac{d^2y}{dt^2} + \{\alpha+\beta-\gamma+1-(\alpha+\beta+1)t\}\frac{dy}{dt} - \alpha\beta y = 0.$$

Since this equation has solution

$$y = A\,{}_2F_1(\alpha, \beta; \alpha+\beta-\gamma+1; t) + \\ + Bt^{\gamma-\alpha-\beta}\,{}_2F_1(\gamma-\alpha, \gamma-\beta; \gamma-\alpha-\beta+1; t),$$

we deduce that the general solution of the hypergeometric equation in the neighbourhood of the regular singular point $x = 1$ is

$$y = A\,{}_2F_1(\alpha, \beta; \alpha+\beta-\gamma+1; 1-x) + \\ + B(1-x)^{\gamma-\alpha-\beta}\,{}_2F_1(\gamma-\alpha, \gamma-\beta; \gamma-\alpha-\beta+1; 1-x) \quad (52.7)$$

provided that $\alpha+\beta-\gamma$ is not an integer or zero.

§53 The Legendre equation

The differential equation

$$(1-x^2)y'' - 2xy' + n(n+1)y = 0, \quad (53.1)$$

has the finite regular singularities $x = \pm 1$; in order to study the solutions valid near $x = 1$ it is advantageous to make the change of independent variable $x = 1-2t$, which transforms the equation into

$$t(1-t)\frac{d^2y}{dt^2} + (1-2t)\frac{dy}{dt} + n(n+1)y = 0$$

This is just the hypergeometric equation with $\alpha = n+1$, $\beta = -n$, $\gamma = 1$, so from the above analysis we see that if we write

$$\phi(t,\sigma)=c_0t^\sigma\left[1+\sum_{r=1}^{\infty}\frac{(\sigma+n+1)_r(\sigma-n)_r}{(\sigma+1)_r(\sigma+1)_r}t^r\right], \tag{53.2}$$

$$\mathbf{L}=t(1-t)\frac{d^2}{dt^2}+(1-2t)\frac{d}{dt}+n(n+1),$$

then

$$\mathbf{L}\phi(t,\sigma)=\sigma^2c_0t^{\sigma-1}.$$

In this instance we have equal roots of the indicial equation: $\sigma=0$ (twice). Obviously one solution is given by

$$\phi(t,0)=c_0\cdot{}_2F_1(n+1,-n;1;t),$$

or, reverting to the original variables in (53.1)

$$c_0P_n(x)=c_0\cdot{}_2F_1\left(n+1,-n;1;\frac{1-x}{2}\right)$$

as found previously in equation (50.7).

Suppose that in the general case the roots of the indicial equation are σ_1,σ_2. Then the function $\phi(x,\sigma_1)$ is a solution as also is the function

$$\frac{\phi(x,\sigma_1)-\phi(x,\sigma_2)}{\sigma_1-\sigma_2}.$$

In the case of equal roots it appears reasonable, therefore, that the required solutions are

$$y_1(x)=\phi(x,\sigma_1)$$

and

$$y_2(x)=\frac{\partial\phi(x,\sigma_1)}{\partial\sigma_1}.$$

This is in fact easily justified and since

$$\frac{\partial x^\sigma}{\partial\sigma}=x^\sigma\log x$$

we see that the two solutions are of the form

$$y_1(x)=(x-a)^\sigma\sum_{r=0}^{\infty}c_r(x-a)^r, \tag{53.3}$$

$$y_2(x)=y_1(x)\log(x-a)+(x-a)^{\sigma_1}\sum_{r=0}^{\infty}\left(\frac{\partial c_r}{\partial \sigma}\right)_{\sigma=\sigma_1}(x-a)^r. \quad (53.4)$$

From this last equation and the expression (53.2) we can find the second solution of Legendre's equation; it is immediately obvious that it has a logarithmic singularity at $x=+1$ (and similarly at $x=-1$).

The hypergeometric function ${}_2F_1(\nu+1, -\nu; 1; \frac{1}{2}-\frac{1}{2}x)$ is called *a Legendre function of the first kind.* We shall confine our attention to the important case when $\nu=n$, a positive integer; since the term in t^{n+1} and every succeeding term in the hypergeometric series ${}_2F_1(a, b; c; t)$ contains the factor $b+n$, it is clear that ${}_2F_1(n+1, -n; 1; t)$ is a polynomial in t of degree n. Hence

$$P_n(x)=1-\frac{n(n+1)}{1}\cdot\frac{1-x}{2}+\frac{(n-1)n(n+1)(n+2)}{(2!)^2}\left(\frac{1-x}{2}\right)^2+\cdots$$

$$+(-1)^n\frac{1\cdot 2\cdot\cdots\cdot(n-1)n(n+1)\cdots 2n}{(n!)^2}\left(\frac{1-x}{2}\right)^n \quad (53.5)$$

is a polynomial in x of degree n such that $P_n(1)=1$. In particular,

$$P_0(x)=1, \quad P_1(x)=x, \quad P_2(x)=\tfrac{1}{2}(3x^2-1), \quad P_3(x)=\tfrac{1}{2}(5x^3-3x).$$

The final term in the above expression for $P_n(x)$ is the only term that gives rise to x^n; we see therefore that the coefficient of x^n is $(2n)!/2^n(n!)^2$. The whole expression may be expanded and then rearranged in descending powers of x, but it is less laborious to substitute the polynomial

$$y=\sum_{r=1}^{n} c_r x^r$$

in the left-hand member of equation (53.1) and to equate to zero the coefficient of each power of x. Taking the coefficient of x^{r-2} we have

$$r(r-1)c_r=(r+n-1)(r-n-2)c_{r-2}.$$

As c_n is known, we work backwards and find

$$c_{n-2}=-\frac{n(n-1)}{2(2n-1)}c_n, \qquad c_{n-4}=\frac{n(n-1)(n-2)(n-3)}{2\cdot 4(2n-1)(2n-3)}c_n$$

$$c_{n-2r}=(-1)^r\frac{n(n-1)\cdots(n-2r+1)}{2^r r!(2n-1)(2n-3)\cdots(2n-2r+1)}c_n$$

It is easily verified that the coefficient of x^{n-1} in the expression of $P_n(x)$ is zero, from which it follows in succession that $c_{n-3}, c_{n-5}, \ldots$ are zero.

Hence we have

$$P_n(x) = \frac{(2n)!}{2^n(n!)^2} \times$$
$$\times \left[x^n - \frac{n(n-1)}{2(2n-1)} x^{n-2} + \frac{n(n-1)(n-2)(n-3)}{2 \cdot 4 \cdot (2n-1)(2n-3)} x^{n-4} - \dots \right]$$

which can be written

$$P_n(x) = \sum_{r=0}^{N} (-1)^r \frac{(2n-2r)!}{2^n r!\,(n-r)!\,(n-2r)!} x^{n-2r} \tag{53.6}$$

where

$$N = \begin{cases} \frac{1}{2}n & \text{for } n \text{ even,} \\ \frac{1}{2}(n-1) & \text{for } n \text{ odd.} \end{cases}$$

Thus the Legendre polynomial $P_n(x)$ is even or odd according as its degree is even or odd; since $P_n(1)=1$, we conclude that $P_n(-1)=(-1)^n$.

We can express equation (53.6) in a different form. Making use of the duplication formula for the gamma function we find that if $r<N$

$$\frac{(2n-2r)!}{2^n(n-r)!\,(n-2r)!} = \frac{\Gamma(n-r+\frac{1}{2})}{r!\,\Gamma(\frac{1}{2}n-r+\frac{1}{2})\Gamma(\frac{1}{2}n-r+1)}.$$

Now using the formula

$$\Gamma(\alpha - r) = (-1)^r \frac{\Gamma(\alpha)}{(1-\alpha)_r}$$

we deduce that

$$(-1)^r \frac{(2n-2r)!}{2^n r!\,(n-r)!\,(n-2r)!} = \frac{(\frac{1}{2}-\frac{1}{2}n)_r(-\frac{1}{2}n)_r}{(\frac{1}{2}-n)_r r!} \cdot \frac{2^n \cdot (\frac{1}{2})_n}{n!}$$

and hence that

$$P_n(x) = \frac{2^n(\frac{1}{2})_n x^n}{n!} \, {}_2F_1(\tfrac{1}{2}-\tfrac{1}{2}n, -\tfrac{1}{2}n; \tfrac{1}{2}-n; x^{-2}) \tag{53.7}$$

Note. An important formula for $P_n(x)$ may be deduced from equation (53.6). We observe that

$$\frac{(2n-2r)!}{(n-2r)!} x^{n-2r} = \frac{d^n}{dx^n} x^{2n-2r}$$

and therefore

$$P_n(x) = \frac{\mathrm{d}^n}{\mathrm{d}x^n} \sum_{r=0}^{N} \frac{(1)^r x^{2n-2r}}{2^n r!\,(n-r)!}$$

$$= \frac{\mathrm{d}^n}{\mathrm{d}x^n} \sum_{r=0}^{n} \binom{n}{r} \frac{(1)^r x^{2n-2r}}{2^n n!}.$$

It will be noticed that under the summation sign r ranges originally from 0 to N. We may however extend the range of summation from $r=0$ to $r=\mathrm{n}$, for the new terms of the sum are of degree less than n and their nth derivatives vanish. Thus we can employ the binomial theorem to obtain *Rodrigues' formula*

$$P_n(x) = \frac{1}{2^n n!} \frac{\mathrm{d}^n}{\mathrm{d}x^n} (x^2-1)^n. \tag{53.8}$$

We may prove from this formula that $P_n(x)$ has n distinct zeros between -1 and $+1$. Let $f_0 = (x^2-1)^n$, $f_m = f'_{m-1}$ for $m = 1, 2, \ldots, n$, so that $f_n = 2^n n!\, P_n(x)$. Then we prove by induction that f_m has at least m distinct zeros in $(-1, 1)$. For this is true of f_0; further, f_{m-1} has zeros at -1 and 1 and so, if it has at least $m-1$ distinct zeros in $(-1, 1)$, then by Rolle's theorem f_m has at least m distinct zeros there. The induction now proceeds.

§54 Solution for large values of $|x|$

Physical problems not infrequently demand the solution of a linear differential equation for large positive or large negative values of the argument x. In such cases we use the reciprocal transformation $x = 1/t$, the differential equation retains its linear form and the point $x = \infty$ is said to be ordinary or singular according as $t=0$ is an ordinary or singular point of the transformed equation. For instance, Legendre's equation is transformed by $x = 1/t$ to the equation

$$(1-t^{-2})\left[t^4 \frac{\mathrm{d}^2 y}{\mathrm{d}t^2} + 2t^3 \frac{\mathrm{d}y}{\mathrm{d}t}\right] - 2t^{-1}\left[t^2 \frac{\mathrm{d}y}{\mathrm{d}t}\right] + n(n+1)y = 0.$$

If we define the operator L_t by the equation

$$\mathsf{L}_t = t^2(t^2-1)\frac{\mathrm{d}^2}{\mathrm{d}t^2} + 2t^3 \frac{\mathrm{d}y}{\mathrm{d}t} + n(n+1), \tag{54.1}$$

this can be written

$$\mathsf{L}_t y = 0. \tag{54.2}$$

Assuming

$$y = \sum_{r=0} c_r t^{\sigma + r},$$

we find that

$$\mathsf{L}_t y = c_0(\sigma + n)(\sigma - n - 1)t^{\sigma},$$

provided that $c_1, c_2, \ldots$ *are so chosen that*

$$(\sigma + r - n + 1)(\sigma + r + n + 2)c_{r+2} = (\sigma + r)(\sigma + r + 1)c_r \qquad (r \geqslant 0).$$

The equation is thus satisfied if $\sigma = -n$ or $n+1$; when $\sigma = -n$ the solution is of the form

$$y = c_0 t^{-n} + c_2 t^{-n+2} + c_4 t^{-n+4} + \cdots$$

with the recurrence relation

$$(r+2)(2n - r - 1)c_{r+2} = -(n-r)(n-r-1)c_r \qquad (r \geqslant 0).$$

Reverting to the variable x we have the first solution

$$y_1(x) = c_0 \left[x^n - \frac{n(n-1)}{2(2n-1)} x^{n-2} + \frac{n(n-1)(n-2)(n-3)}{2 \cdot 4(2n-1)(2n-3)} x^{n-4} - \cdots \right]$$

and taking $c_0 = (2n)!/2^n(n!)^2$ when n is a positive integer we identify this solution with the polynomial $P_n(x)$.

When $\sigma = n+1$ the solution is

$$y = c_0 t^{n+1} + c_2 t^{n+3} + c_4 t^{n+5} + \cdots$$

with the recurrence relation

$$(r+2)(2n + r + 3)c_{r+2} = (n + r + 1)(n + r + 2)c_r.$$

In this way we obtain the second solution

$$y_2(x) = c_0 \left[x^{-n-1} + \frac{(n+1)(n+2)}{2(2n+3)} x^{-n-3} + \right.$$

$$\left. + \frac{(n+1)(n+2)(n+3)(n+4)}{2 \cdot 4 \cdot (2n+3)(2n+5)} x^{-n-5} + \cdots \right].$$

When n is a positive integer this series does not terminate, but since

$$\frac{c_{r+2}}{c_r}=\frac{(n+r+1)(n+r+2)}{(r+2)(2n+2r+1)}\to 1 \qquad \text{as } r\to\infty$$

it converges for $|x|>1$. Taking $c_0=2^n(n!)^2/(2n+1)!$ we obtain the second solution

$$Q_n(x)=\sum_{r=0}^{\infty}\frac{2^n(n+r)!\,(n+2r)!}{r!\,(2n+2r+1)!}x^{-n-2r-1}$$

which is known as the *Legendre function of the second kind.*

It can be written in the alternative form

$$Q_n(x)=\frac{x^{-n-1}}{2^n(\frac{3}{2})_n}\,{}_2F_1(\tfrac{1}{2}n+\tfrac{1}{2},\tfrac{1}{2}n+1;n+\tfrac{3}{2};x^{-2}). \tag{54.3}$$

We have already used in equation (50.11) the symbol $Q_n(x)$ for a second solution of Legendre's equation. It is possible to prove this expression is equal to (54.3), provided that the lower limit is taken at infinity, i.e.

$$Q_n(x)=P_n(x)\int_{\infty}^{x}\frac{dt}{(1-t^2)\{P_n(t)\}^2} \tag{54.4}$$

§55 The Bessel equation and the function $J_n(x)$

Unlike the Legendre equation, the Bessel equation

$$x^2y''+xy'+(x^2-n^2)y=0. \tag{55.1}$$

is not essentially a particular case of the hypergeometric equation. It has only the one finite singularity $x=0$, which is regular; the point at infinity is also a singularity, but is irregular.

To obtain a solution in series by the Frobenius method, write $\mathbf{L}y$ for the left-hand member of the equation and

$$\phi(x,\sigma)=\sum_{r=0}^{\infty}c_rx^{\sigma+r} \qquad (c_0\neq 0).$$

Then

$$\mathbf{L}\phi(x,\sigma)=c_0(\sigma^2-n^2)x^{\sigma}+c_1\{(\sigma+1)^2-n^2\}x^{\sigma+1},$$

provided the coefficients $c_1, c_2, \ldots, c_r, \ldots$ are such that

$$\{(\sigma+r)^2-n^2\}c_r+c_{r-2}=0 \qquad (r>2).$$

If $y=\phi(x,\sigma)$ is a solution, $\mathbf{L}\phi(x,\sigma)$ must vanish identically, and therefore

$$c_0(\sigma^2-n^2)=0, \qquad c_1\{(\sigma+1)^2-n^2\}=0.$$

Since $c_0 \neq 0$ we must take $\sigma=\pm n$ and consequently $c_1=0$. The relation between c_r and c_{r-2} now shows, taking $r=3, 5, \ldots$ in succession, that all coefficients of odd rank vanish.

Taking first of all $\sigma=n$ and writing $r=2s$ we have

$$4s(n+s)c_{2s}=-c_{2s-2} \qquad (s>1)$$

and thus we obtain the solution

$$y=\phi(x,n)=c_0\left\{x^n-\frac{x^{n+2}}{2^2(n+1)}+\cdots+\frac{(-)^s x^{n+2s}}{2^{2s}(n+1)\cdots(n+s)\cdot s!}+\cdots\right\}.$$

The coefficients are finite except when n is a negative integer. Excluding this case, we standardize the solution by taking $c_0=1/2^n\Gamma(n+1)$ in general and $c_0=1/2^n n!$ when n is a positive integer. Since

$$\frac{c_{2s}}{c_{2s-2}}=-\frac{1}{4s(n+s)}\to 0 \qquad \text{as } s\to\infty,$$

the series converges for any finite value of x, and thus we have the first solution $y=J_n(x)$, where

$$J_n(x)=\sum_{s=0}^{\infty}\frac{(-)^s}{s!\,\Gamma(n+s+1)}\left(\frac{x}{2}\right)^{n+2s}$$

in general. In particular, when n is a positive integer,

$$J_n(x)=\sum_{s=0}^{\infty}\frac{(-)^s}{s!\,(n+s)!}\left(\frac{x}{2}\right)^{n+2s}$$

This function $J_n(x)$ is known as *the Bessel function of the first kind of order n*.

When n is **not** an integer the second solution may be obtained by replacing n by $-n$; it is therefore

$$J_{-n}(x)=\sum_{s=0}^{\infty}\frac{(-)^s}{s!\,\Gamma(s-n+1)}\left(\frac{x}{2}\right)^{2s-n}$$

The leading terms of $J_n(x)$ and $J_{-n}(x)$ are respectively finite (non-zero) multiples of n^n and x^{-n}; the one is not a mere multiple of the other and therefore the general solution of the Bessel equation may be expressed as

$$y=AJ_n(x)+BJ_{-n}(x).$$

But when n is an integer, and since n appears in the differential equation only as n^2, there is no loss of generality in taking it to be a positive integer; $J_{-n}(x)$ is not distinct from $J_n(x)$. For the denominators of the first n terms of the series for $J_{-n}(x)$ contain respectively the factors $\Gamma(1-n), \Gamma(2-n), \ldots, \Gamma(0)$ which are infinite, and so these terms vanish. Thus

$$J_{-n}(x)=\sum_{s=n}^{\infty}\frac{(-)^s}{s!\,\Gamma(s-n+1)}\left(\frac{x}{2}\right)^{2s-n}$$

$$=\sum_{r=0}^{\infty}\frac{(-)^{r+n}}{(r+n)!\,\Gamma(r+1)}\left(\frac{x}{2}\right)^{2r+n}=(-)^n J_n(x).$$

In this case a modification of the method of Frobenius is necessary to give an independent second solution.

§56 The function $Y_n(x)$

Consider the functions

$$J_{n+\varepsilon}(x)=\sum_{s=0}^{\infty}\frac{(-)^s}{s!\,\Gamma(s+n+\varepsilon+1)}\left(\frac{x}{2}\right)^{2s+n+\varepsilon},$$

$$J_{-n-\varepsilon}(x)=\sum_{s=0}^{\infty}\frac{(-)^s}{s!\,\Gamma(s-n-\varepsilon+1)}\left(\frac{x}{2}\right)^{2s-n-\varepsilon},$$

where n is a positive integer and ε a real number numerically less than unity. Both satisfy the Bessel equation

$$x^2y''+xy'+\{x^2-(n+\varepsilon)^2\}y=0$$

and so does the function

$$\eta=\frac{J_{n+\varepsilon}(x)-(-)^n J_{-n-\varepsilon}(x)}{\varepsilon}.$$

So if $\mathbf{L}y\equiv x^2y''+xy'+(x^2-n^2)y$, then

$$\mathbf{L}\eta=\{(n+\varepsilon)^2-n^2\}\eta=\varepsilon(2n+\varepsilon)\eta.$$

If, therefore, η has a limit as $\varepsilon\to 0$, $y=\lim\eta$ is a solution of $\mathbf{L}y=0$, i.e. is a Bessel function of order n. We prove that the limit exists by making the following transformation:

$$\lim_{\varepsilon\to 0}\frac{J_{n+\varepsilon}(x)-(-)^n J_{-n-\varepsilon}(x)}{\varepsilon}=$$

$$=\lim_{\varepsilon\to 0}\left\{\frac{J_{n+\varepsilon}(x)-J_n(x)}{\varepsilon}+(-)^n\frac{J_{-n}(x)-J_{-n-\varepsilon}(x)}{\varepsilon}\right\}$$

$$=\left[\frac{\partial J_\nu(x)}{\partial\nu}-(-)^n\frac{\partial J_{-\nu}(x)}{\partial\nu}\right]_{\nu=n}$$

and evaluating these partial derivatives. Some authorities take the latter expression to be the definition of the standard second solution $Y_n(x)$, but we shall follow the modern practice of introducing the factor π^{-1} and defining $Y_n(x)$ thus†

$$Y_n(x)=\frac{1}{\pi}\left[\frac{\partial J_\nu(x)}{\partial\nu}-(-)^n\frac{\partial J_{-\nu}(x)}{\partial\nu}\right]_{\nu=n}.$$

Since

$$J_\nu(x)=\sum_{s=0}^{\infty}\frac{(-)^s}{s!\,\Gamma(s+\nu+1)}\left(\frac{x}{2}\right)^{2s+\nu}$$

we have

$$\frac{\partial J_\nu(x)}{\partial\nu}=\sum_{s=0}^{\infty}\frac{(-)^s}{s!\,\Gamma(s+\nu+1)}\left(\frac{x}{2}\right)^{2s+\nu}\{\log(\tfrac{1}{2}x)-\psi(s+\nu+1)\},$$

where $\psi(z)=\Gamma'(z)/\Gamma(z)$. Hence‡

$$\left[\frac{\partial J_\nu(x)}{\partial\nu}\right]_{\nu=n}=\sum_{s=0}^{\infty}\frac{(-)^s}{s!\,(s+n)!}\left(\frac{x}{2}\right)^{2s+n}\times$$

$$\times\left\{\log(\tfrac{1}{2}x)+\gamma-1-\tfrac{1}{2}-\cdots-\frac{1}{s+n}\right\}=$$

† This definition was adopted in the *British Association Mathematical Tables*, vol. vi (Cambridge, 1937). An equivalent definition is

$$Y_n(x)=\lim_{\nu\to n}\frac{J_\nu(x)\cos\nu\pi-J_{-\nu}(x)}{\sin\nu\pi}.$$

‡ Since $\Gamma(z+1)=z\Gamma(z)$, we have $\psi(z+1)=\psi(z)+1/z$ and so when z is a positive integer,

$$\psi(z+1)=\psi(1)+1+\frac{1}{2}+\frac{1}{3}+\cdots+\frac{1}{z}$$

where $-\psi(1)=\gamma=0.57721566\ldots$ (Euler's constant).

$$=J_n(x)\{\log(\tfrac{1}{2}x)+\gamma\}-\sum_{s=0}^{\infty}\frac{(-)^s}{s!\,(s+n)!}\times$$

$$\times\left(1+\tfrac{1}{2}+\cdots+\frac{1}{s+n}\right)\left(\frac{x}{2}\right)^{2s+n}.$$

It will be remembered that the first n terms in $J_{-\nu}(x)$ vanish when $\nu=n$; we therefore write

$$J_{-\nu}(x)=\sum_{s=0}^{n-1}\frac{(-)^s}{s!\,\Gamma(s-\nu+1)}\left(\frac{x}{2}\right)^{2s-\nu}+\sum_{s=n}^{\infty}\frac{(-)^s}{s!\,\Gamma(s-\nu+1)}\left(\frac{x}{2}\right)^{2s-\nu}.$$

In the first sum we use the relation

$$\Gamma(z)\Gamma(1-z)=\pi/\sin z\pi \qquad \text{with } z=s-\nu+1;$$

in the second we replace s by $s+n$, and thus obtain

$$J_{-\nu}(x)=\sum_{s=0}^{n-1}\frac{(-)^s\sin(s-\nu+1)\pi\cdot\Gamma(\nu-s)}{\pi\cdot s!}\left(\frac{x}{2}\right)^{2s-\nu}+$$

$$+\sum_{s=0}^{\infty}\frac{(-)^{s+n}}{(s+n)!\,\Gamma(s+n-\nu+1)}\left(\frac{x}{2}\right)^{2s+2n-\nu}$$

Differentiating with respect to ν, we have

$$\frac{\partial J_{-\nu}(x)}{\partial\nu}=\sum_{s=0}^{n-1}\frac{(-)^s}{\pi\cdot s!}\left(\frac{x}{2}\right)^{2s-\nu}\sin(s-\nu+1)\pi\cdot\Gamma(\nu-s)\times$$

$$\times\left\{-\frac{\pi\cos(s-\nu+1)\pi}{\sin(s-\nu+1)\pi}+\psi(\nu-s)-\log\left(\frac{x}{2}\right)\right\}+$$

$$+\sum_{s=0}^{\infty}\frac{(-)^{s+n}}{(s+n)!\,\Gamma(s+n-\nu+1)}\left(\frac{x}{2}\right)^{2s+2n-\nu}\times$$

$$\times\left\{\psi(s+n-\nu+1)-\log\left(\frac{x}{2}\right)\right\}$$

$$\left[\frac{\partial J_{-\nu}(x)}{\partial\nu}\right]_{\nu=n}=\sum_{s=0}^{n-1}\frac{(-)^s}{s!}\left(\frac{x}{2}\right)^{2s-n}\Gamma(n-s)\cos(s-n)\pi+$$

$$+\sum_{s=0}^{\infty}\frac{(-)^{s+n}}{(s+n)!\,s!}\left(\frac{x}{2}\right)^{2s+n}\left\{\psi(s+1)-\log\left(\frac{x}{2}\right)\right\}=$$

$$=(-)^n\sum_{s=0}^{n-1}\frac{(n-s-1)!}{s!}\left(\frac{x}{2}\right)^{2s-n}-\sum_{s=0}^{\infty}\frac{(-)^{s-n}}{s!\,(s+n)!}\left(\frac{x}{2}\right)^{2s+n}\times$$

$$\times\left\{\log\left(\frac{x}{2}\right)+\gamma-1-\frac{1}{2}-\cdots-\frac{1}{s}\right\}=$$

$$=(-)^n\sum_{s=0}^{n-1}\frac{(n-s-1)!}{s!}\left(\frac{x}{2}\right)^{2s-n}-(-)^nJ_n(x)\{\log(\tfrac{1}{2}x)+\gamma\}+$$

$$+(-1)^n\sum_{s=0}^{\infty}\frac{(-)^s}{s!\,(s+n)!}\left(1+\frac{1}{2}+\cdots+\frac{1}{s}\right)\left(\frac{x}{2}\right)^{2s+n}.$$

Thus the function $Y_n(x)$ or *Bessel function of the second kind of order n* is given by

$$\pi Y_n(x)=\left[\frac{\partial J_\nu(x)}{\partial\nu}\right]_{\nu=n}-(-)^n\left[\frac{\partial J_{-\nu}(x)}{\partial\nu}\right]_{\nu=n}=$$

$$=2J_n(x)\{\log(\tfrac{1}{2}x)+\gamma\}-\sum_{s=0}^{n-1}\frac{(n-s-1)!}{s!}\left(\frac{x}{2}\right)^{2s-n}-$$

$$-\sum_{s=0}^{\infty}\frac{(-)^s}{s!\,(n+s)!}\left(1+\frac{1}{2}+\cdots+\frac{1}{s}+1+\frac{1}{2}+\cdots+\frac{1}{n+s}\right)\left(\frac{x}{2}\right)^{n+2s}$$

and when n is a positive integer the complete solution of the Bessel equation is

$$y=AJ_n(x)+BY_n(x).$$

7

Linear systems

§57 Existence of solutions

We consider now the vector equation

$$\dot{\mathbf{y}} = \mathbf{A}\mathbf{y} \tag{57.1}$$

where

$$\mathbf{A} = \mathbf{A}(t) = [a_{ij}(t)] \tag{57.2}$$

with $a_{ij}(t) \in C[a, b]$. The region on which the right-hand side of (57.1) is defined is

$$\Omega = \{(t, \mathbf{y}) : a < t < b, \|\mathbf{y}\| < \infty\}.$$

Since $a_{ij}(t) \in C[a, b]$, they are bounded. Hence the norm of $\mathbf{A}$ relative to $[a, b]$ exists and may be denoted by some real $M > 0$. It follows that

$$\|\mathbf{A}(t)\| \leqslant M, \ \forall t \in [a, b].$$

Also if $(t, \mathbf{y}^1) \in \Omega$, $(t, \mathbf{y}^2) \in \Omega$,

$$\|\mathbf{A}(t)\mathbf{y}^1 - \mathbf{A}(t)\mathbf{y}^2\| \leqslant \|\mathbf{A}(t)\| \cdot \|\mathbf{y}^1 - \mathbf{y}^2\| \leqslant M\|\mathbf{y}^1 - \mathbf{y}^2\|,$$

showing that the equation satisfies a Lipschitz condition in Ω.

It follows that the initial value problem

$$\dot{\mathbf{y}} = \mathbf{A}\mathbf{y}, \qquad \mathbf{y}(t_0) = \mathbf{y}^0 \tag{57.3}$$

has a unique solution for any $t_0 \in [a, b]$, $\mathbf{y}^0 \in \mathbb{R}^n$, and by an application of the continuation process this solution can be shown to be valid on the fundamental interval $[a, b]$.

It should be recalled that

(i) this solution is a continuous function of the initial conditions;
(ii) if A depends continuously on a set of parameters, the solution is a continuous function of these parameters.

§58 The solution space

Theorem 58.1 *The solutions of the linear differential equation*

$$\dot{\mathbf{y}} = \mathbf{A}(t)\mathbf{y}$$

form a vector space.

Proof. Suppose that $\mathbf{y}^1, \mathbf{y}^2$ are solutions of (57.1) and that c_1 and c_2 are real (arbitrary) constants. Then

$$\frac{\mathrm{d}}{\mathrm{d}t}(c_1\mathbf{y}^1 + c_2\mathbf{y}^2) = c_1\dot{\mathbf{y}}^1 + c_2\dot{\mathbf{y}}^2 = c_1\mathbf{A}(t)\mathbf{y}^1 + c_2\mathbf{A}(t)\mathbf{y}^2$$

$$= \mathbf{A}(t)(c_1\mathbf{y}^1 + c_2\mathbf{y}^2)$$

showing that $c_1\mathbf{y}^1 + c_2\mathbf{y}^2$ is a solution of (57.1).

The solutions, therefore, form a vector space which is a subspace of all n-vectors whose components belong to the class $C[a, b]$.

We shall call this vector space the *solution space* of the equation (57.1). That the solution space has dimension n is proved in:

Theorem 58.2 *The equation* $\dot{\mathbf{y}} = \mathbf{A}(t)\mathbf{y}$ *has n linearly independent solutions and every solution of the equation is expressible as a linear combination of these n solutions.*

Proof. Let $t_0 \in [a, b]$ and let $\mathbf{b}^1, \ldots, \mathbf{b}^n$ be any set of linearly independent n vectors in $\mathbb{R}^n$.

Suppose now that $\mathbf{u}^j(t)$ is the solution of the initial-value problem

$$\dot{\mathbf{y}} = \mathbf{A}(t)\mathbf{y}, \quad \mathbf{y}(t_0) = \mathbf{b}^j, \qquad (j = 1, \ldots, n).$$

If the set $\{\mathbf{u}^j(t)\}$ were linearly dependent there would exist scalars $c_1, \ldots, c_n$, not all zero, such that

$$\sum_{j=1}^{n} c_j\mathbf{u}^j(t_0) = \mathbf{0}.$$

This would imply that

$$\sum_{j=1}^{n} c_j\mathbf{b}^j = \mathbf{0}$$

in contradiction to our hypothesis.

Hence $\mathbf{u}^j(t)$ is a linearly independent set.

To prove the second part of the theorem, let $\mathbf{u}(t)$ be any solution of equation (57.1). Since the set $\{\mathbf{b}^j\}$ is a basis for $\mathbb{R}^n$, there exists a set of scalars $c_1, \ldots, c_n$ such that

$$\mathbf{u}(t_0) = \sum_{j=1}^{n} c_j \mathbf{b}^j.$$

Having determined the values of these scalars we define a vector $\mathbf{v}(t)$ by the equation

$$\mathbf{v}(t) = \sum_{j=1}^{n} c_j \mathbf{u}^j(t).$$

Clearly $\mathbf{v}(t)$ is a solution of (57.1) such that

$$\mathbf{v}(t_0) = \mathbf{u}(t_0).$$

The uniqueness property guarantees that

$$\mathbf{v}(t) = \mathbf{u}(t)$$

i.e. that

$$\mathbf{u}(t) = \sum_{j=1}^{n} c_j \mathbf{u}^j(t).$$

This proves the theorem.

Corollary 58.2a *Any set of n linearly independent solutions of* $\dot{\mathbf{y}} = \mathbf{A}\mathbf{y}$ *is a basis for its solution space.*

Corollary 58.2b *If all the components of a solution* $\mathbf{u}$ *of* $\dot{\mathbf{y}} = \mathbf{A}\mathbf{y}$ *are zero at some point* $t_0 \in [a, b]$ *then* $\mathbf{u}(t) \equiv \mathbf{0}$.

Corollary 58.2c *If a set of solutions* $u_1, u_2, \ldots, u_k$ *is such that their functional values* $u_1(t_0), \ldots, u_k(t_0)$ *form a linearly dependent set for any* $t_0 \in [a, b]$, *then the solutions are linearly dependent at every* $t \in [a, b]$.

Definition Any set of n linearly independent solutions of $\dot{\mathbf{y}} = \mathbf{A}(t)\mathbf{y}$ is called a *fundamental set of solutions* of the equation.

The matrix with these solutions as columns is called a *fundamental matrix* of the given equation.

Suppose that $\mathbf{y}^1, \mathbf{y}^2, \ldots, \mathbf{y}^n$ form a fundamental set of solutions of equation (57.1) and let

$$\mathbf{Y} = [\mathbf{y}^1, \mathbf{y}^2, \ldots, \mathbf{y}^n]$$

be the corresponding fundamental matrix.

Clearly, if $\mathbf{c}$ is any constant n-vector, $\mathbf{Yc}$ is a solution of (57.1), and every solution of that equation can be expressed in this form. From Corollary 58.2c we see that $\mathbf{Y}(t)$ is non-singular at each point $t \in [a, b]$, so that its inverse $\mathbf{Y}^{-1}(t)$ is defined on $[a, b]$, and its determinant $|\mathbf{Y}(t)|$ is not zero, for any $t \in [a, b]$. In addition, if $\mathbf{C}$ is any constant non-singular matrix of order n, then $\mathbf{YC}$ is a fundamental matrix and every fundamental matrix can be expressed in this form.

The matrix equation

$$\dot{\mathbf{Y}} = \mathbf{AY} \tag{58.1}$$

can be regarded as a generalization of the vector equation

$$\dot{\mathbf{y}} = \mathbf{Ay}. \tag{58.2}$$

It is easily seen that a matrix is a solution of (58.1) if and only if its columns regarded as vectors are solutions of (58.2). A fundamental matrix for (58.2) will also be called a fundamental matrix for (58.1). The solutions of the matrix equation are given by $\mathbf{YC}$, where $\mathbf{C}$ is an arbitrary constant matrix with n rows and columns.

Theorem 58.3 *A vector (or matrix) equation is completely determined by one of its fundamental matrices.*

Proof. If $\mathbf{Y}$ is a fundamental matrix for

$$\dot{\mathbf{y}} = \mathbf{Ay},$$

then

$$\dot{\mathbf{Y}} = \mathbf{AY}.$$

It follows that

$$\mathbf{A} = \dot{\mathbf{Y}} \cdot \mathbf{Y}^{-1}.$$

Since the equation is determined by the matrix $\mathbf{A}$, the proof of the theorem is complete.

Definition The *trace* of an n by n matrix $\mathbf{B}$, designated by $tr\,\mathbf{B}$, is defined by

$$tr\,\mathbf{B} = \sum_{j=1}^{n} b_{jj}.$$

We know from linear algebra that the trace and the determinant of a matrix are invariant under a similarity transformation.

Theorem 58.4 *If* $\mathbf{Y}$ *is a fundamental matrix for the equation* $\dot{\mathbf{Y}}=\mathbf{AY}$ *and if* $|\mathbf{Y}|$ *denotes the determinant of* $\mathbf{Y}$, *then for all* $t_0 \in [a, b]$

$$|\mathbf{Y}(t)| = |\mathbf{Y}(t_0)| \exp\left[\int_{t_0}^{t} tr\, \mathbf{A}(s)\, \mathrm{d}s\right].$$

Proof. Let

$$\mathbf{Y} = \begin{vmatrix} y_1^1 & y_1^2 & \cdots & y_1^n \\ y_2^1 & y_2^2 & \cdots & y_2^n \\ \vdots & \vdots & & \vdots \\ y_n^1 & y_n^2 & \cdots & y_n^n \end{vmatrix}$$

where y_j^i is the jth component of $\mathbf{y}^i$.

By the rule for differentiating a determinant, we have

$$|\dot{\mathbf{Y}}| = \begin{vmatrix} \dot{y}_1^1 & \dot{y}_1^2 & \cdots & \dot{y}_1^n \\ y_2^1 & y_2^2 & \cdots & y_2^n \\ \vdots & \vdots & & \vdots \\ y_n^1 & y_n^2 & \cdots & y_n^n \end{vmatrix} + \begin{vmatrix} y_1^1 & y_1^2 & \cdots & y_1^n \\ \dot{y}_2^1 & \dot{y}_2^2 & \cdots & \dot{y}_2^n \\ \vdots & \vdots & & \vdots \\ y_n^1 & y_n^2 & \cdots & y_n^n \end{vmatrix} + \cdots + \begin{vmatrix} y_1^1 & y_1^2 & \cdots & y_1^n \\ y_2^1 & y_2^2 & \cdots & y_2^n \\ \vdots & \vdots & & \vdots \\ \dot{y}_n^1 & \dot{y}_n^2 & \cdots & \dot{y}_n^n \end{vmatrix}$$

We denote by I_{hl} the matrix each of whose elements is zero except for the element in the hth row and the lth column which has the value unity, i.e.

$$I_{hl} = \delta_{ih}\delta_{lj}$$

where δ_{ij} is the Kronecker delta. We also define I^{hh} by

$$I^{hh} = I - I_{hh} \qquad (h = 1, 2, \ldots, n).$$

Using this notation we can write

$$|\dot{\mathbf{Y}}| = \sum_{j=1}^{n} |\mathbf{I}^{jj}\mathbf{Y} + \mathbf{I}_{jj}\dot{\mathbf{Y}}|.$$

Since $\mathbf{Y}$ is a solution of (58.1) we have

$$|\dot{\mathbf{Y}}| = \sum_{j=1}^{n} |\mathbf{I}^{jj}\mathbf{Y} + \mathbf{I}_{jj}\mathbf{AY}| = \sum_{j=1}^{n} |\mathbf{I}^{jj} + \mathbf{I}_{jj}\mathbf{A}| \cdot |\mathbf{Y}| =$$

$$= \sum_{j=1}^{n} a_{jj}|\mathbf{Y}| = (tr\, \mathbf{A})|\mathbf{Y}|.$$

Since $Y=0$ on $[a,b]$,

$$|\mathbf{Y}(t)|=|\mathbf{Y}(t_0)|\exp\left[\int_{t_0}^{t} tr\,\mathbf{A}(s)\,\mathrm{d}s\right].$$

This proves the theorem.

Example 58.1 Find a fundamental matrix for the equation

$$\dot{\mathbf{y}}=\begin{bmatrix}1 & 1\\ 0 & -1\end{bmatrix}\mathbf{y}.$$

In component form this equation is equivalent to

$$\begin{aligned}\dot{y}_1&=y_1+y_2\\ \dot{y}_2&=\quad -y_2.\end{aligned}$$

The second equation has solution

$$y_2=c_2\,\mathrm{e}^{-t}.$$

Substituting this in the first equation we have

$$\frac{\mathrm{d}}{\mathrm{d}t}(y_1\,\mathrm{e}^{-t})=c_2\,\mathrm{e}^{-2t}$$

which shows that

$$y=c_1\,\mathrm{e}^{+t}-\tfrac{1}{2}c_2\,\mathrm{e}^{-t}.$$

We therefore have

$$\mathbf{y}^1=\begin{bmatrix}\mathrm{e}^t\\ 0\end{bmatrix},\quad \mathbf{y}^2=\begin{bmatrix}-\tfrac{1}{2}\,\mathrm{e}^{-t}\\ \mathrm{e}^{-t}\end{bmatrix}$$

so that

$$\mathbf{Y}=\begin{bmatrix}\mathrm{e}^t & -\tfrac{1}{2}\mathrm{e}^{-t}\\ 0 & \mathrm{e}^{-t}\end{bmatrix}.$$

Example 58.2 Find the vector differential equation which has fundamental matrix

$$\begin{bmatrix}\mathrm{e}^t & t\,\mathrm{e}^t\\ \mathrm{e}^t & (t+1)\,\mathrm{e}^t\end{bmatrix}.$$

Let

$$\mathbf{Y}(t)=\begin{bmatrix} e^t & t\,e^t \\ e^t & (t+1)\,e^t \end{bmatrix};$$

then

$$\mathbf{Y}^{-1}(t)=\begin{bmatrix} (t+1)\,e^{-t} & -t\,e^{-t} \\ -e^{-t} & e^{-t} \end{bmatrix}, \quad \dot{\mathbf{Y}}(t)=\begin{bmatrix} e^t & (t+1)\,e^t \\ e^t & (t+2)\,e^t \end{bmatrix}$$

so that

$$\mathbf{A}(t)=\begin{bmatrix} 1 & t+1 \\ 1 & t+2 \end{bmatrix}\begin{bmatrix} t+1 & -t \\ -1 & 1 \end{bmatrix}$$

i.e.

$$\mathbf{A}(t)=\begin{bmatrix} 0 & 1 \\ -1 & 2 \end{bmatrix}.$$

§59 The non-homogeneous equation

The equation

$$\dot{\mathbf{y}}=\mathbf{A}\mathbf{y}+\mathbf{f} \tag{59.1}$$

where $\mathbf{f}=(f_1,\ldots,f_n)$, $f_j\in C[a,b]$ is commonly called *inhomogeneous* whereas the form with $\mathbf{f}=0$

$$\dot{\mathbf{y}}=\mathbf{A}\mathbf{y} \tag{59.2}$$

is called *homogeneous.*

We see that the solutions of (59.1) do not form a vector space (unless $\mathbf{f}=\mathbf{0}$) but that they can be simply expressed in terms of the solutions of the homogeneous equation. This is shown in:

Theorem 59.1 *If $\bar{\mathbf{y}}$ is any particular solution of the non-homogeneous equation $\dot{\mathbf{y}}=\mathbf{A}\mathbf{y}+\mathbf{f}$ and $\mathbf{Y}$ is a fundamental matrix for the corresponding homogeneous equation $\dot{\mathbf{y}}=\mathbf{A}\mathbf{y}$, then*

$$\mathbf{y}=\bar{\mathbf{y}}+\mathbf{Y}\mathbf{c}$$

is a solution of $\dot{\mathbf{y}}=\mathbf{A}\mathbf{y}+\mathbf{f}$ for every $\mathbf{c}\in\mathbb{R}^n$ and every solution is of this form.

Proof. The direct result is easily verified.

For the converse, suppose that $\mathbf{u}$ is any solution of (59.1) then,

$$\frac{d}{dt}(\mathbf{u}-\bar{\mathbf{y}})=\mathbf{A}(\mathbf{u}-\bar{\mathbf{y}}).$$

Hence $\mathbf{u}-\mathbf{y}$ is a solution of (59.2). It follows that, for some constant vector $\mathbf{c}$,

$$\mathbf{u}-\bar{\mathbf{y}}=\mathbf{Y}\mathbf{c}.$$

Hence

$$\mathbf{u}=\bar{\mathbf{y}}+\mathbf{Y}\mathbf{c} \tag{59.3}$$

which proves the theorem.

To find $\bar{\mathbf{y}}(t)$ we may use:

Theorem 59.2 *The function*

$$\mathbf{Y}(t)\int_a^t \mathbf{Y}^{-1}(s)\mathbf{f}(s)\,\mathrm{d}s$$

in which $\mathbf{Y}(t)$ *is a fundamental matrix for* $\dot{\mathbf{y}}=\mathbf{A}\mathbf{y}$ *is a particular solution of* $\dot{\mathbf{y}}=\mathbf{A}\mathbf{y}+\mathbf{f}$.

Proof. Let $\bar{\mathbf{y}}=\mathbf{Y}\mathbf{h}$, then

$$\dot{\bar{\mathbf{y}}}=\dot{\mathbf{Y}}\mathbf{h}+\mathbf{Y}\dot{\mathbf{h}}.$$

But to be a particular solution

$$\dot{\bar{\mathbf{y}}}=\mathbf{A}\mathbf{Y}\mathbf{h}+\mathbf{f}$$

i.e.

$$\mathbf{Y}\dot{\mathbf{h}}=(\mathbf{A}\mathbf{Y}-\dot{\mathbf{Y}})\mathbf{h}+\mathbf{f}.$$

Since $\mathbf{Y}$ is a fundamental matrix, $\mathbf{A}\mathbf{Y}=\dot{\mathbf{Y}}$ so that $\mathbf{Y}\dot{\mathbf{h}}=\mathbf{f}$ and

$$\dot{\mathbf{h}}(t)=\mathbf{Y}^{-1}(t)\mathbf{f}(t).$$

Integrating, we obtain

$$\mathbf{h}(t)=\int_a^t \mathbf{Y}^{-1}(s)\mathbf{f}(s)\,\mathrm{d}s.$$

Hence we have the solution

$$\mathbf{y}(t)=\mathbf{Y}(t)\int_a^t \mathbf{Y}^{-1}(s)\mathbf{f}(s)\,\mathrm{d}s. \tag{59.4}$$

Corollary 59.2 *The matrix equation* $\dot{\mathbf{Y}}=\mathbf{A}\mathbf{Y}+\mathbf{B}$, *where* $\mathbf{B}$ *is any* $n\times k$

matrix with components in $C[a,b]$, has a particular solution

$$\mathbf{Y}(t)\int_a^t \mathbf{Y}^{-1}(s)\mathbf{B}(s)\,ds.$$

Example 59.1 Find a particular solution of the equation

$$\dot{\mathbf{y}}=\begin{bmatrix}1 & 1\\ 0 & -1\end{bmatrix}\mathbf{y}+\begin{bmatrix}1\\ 0\end{bmatrix}.$$

From Example **58.1** we have

$$\mathbf{Y}(t)=\begin{bmatrix}e^t & -\frac{1}{2}e^{-t}\\ 0 & e^{-t}\end{bmatrix},\qquad \mathbf{Y}^{-1}(s)=\begin{bmatrix}e^{-s} & \frac{1}{2}e^{-s}\\ 0 & e^s\end{bmatrix}$$

Taking

$$\mathbf{f}(s)=\begin{bmatrix}1\\ 0\end{bmatrix},$$

we have

$$\mathbf{Y}(t)\int_0^t \mathbf{Y}^{-1}(s)\mathbf{f}(s)\,ds=\begin{bmatrix}e^t & -\frac{1}{2}e^{-t}\\ 0 & e^{-t}\end{bmatrix}\begin{bmatrix}1-e^{-t}\\ 0\end{bmatrix}=\begin{bmatrix}e^t-1\\ 0\end{bmatrix}$$

But

$$\begin{bmatrix}e^t\\ 0\end{bmatrix}$$

is part of the general solution so we may take

$$\begin{bmatrix}-1\\ 0\end{bmatrix}$$

to be a particular solution.

§60 The *n*th-order linear homogeneous equation

We consider the equation

$$\mathsf{L}u=0 \tag{60.1}$$

where

$$\mathsf{L}=D^n+p_1D^{n-1}+\cdots+p_{n-1}D+p_n,\qquad D=d/dt,$$

$$p_j(t)\in C[a,b]. \tag{60.2}$$

Definition A function u is a solution of (60.1) if it belongs to $C^n[a,b]$ and satisfies the equation at each point $t \in [a,b]$.

Equation (60.1) together with the auxiliary conditions

$$\begin{aligned} u(t_0) &= b_1 \\ u'(t_0) &= b_2 \\ &\vdots \\ u^{(n-1)}(t_0) &= b_n, \end{aligned} \tag{60.3}$$

is an initial-value problem.

If u is any function with $n-1$ derivatives, the vector

$$k(u) = \begin{bmatrix} u \\ u' \\ \vdots \\ u^{(n-1)} \end{bmatrix}$$

is called the *Wronskian vector of* u. If $u_1, u_2, \ldots, u_n$ is any set of n functions with $n-1$ derivatives and

$$\mathbf{u} = (u_1, u_2, \ldots, u_n)$$

the matrix

$$\mathbf{K}(\mathbf{u}) = \begin{bmatrix} u_1 & u_2 & \ldots & u_n \\ u_1' & u_2' & \ldots & u_n' \\ \vdots & \vdots & & \vdots \\ u_1^{(n-1)} & u_2^{(n-1)} & \ldots & u_n^{(n-1)} \end{bmatrix}$$

is called the *Wronskian matrix of* $\mathbf{u}$ (or of the components of $\mathbf{u}$). Obviously, the jth column of the Wronskian matrix $\mathbf{K}(\mathbf{u})$ is the Wronskian vector of the component u_j of $\mathbf{u}$, so that

$$\mathbf{K}(\mathbf{u}) = [\mathbf{k}(u_1), \mathbf{k}(u_2), \ldots, \mathbf{k}(u_n)].$$

With this notation we can write equation (60.1) in the form

$$u^{(n)}(t) + \tilde{\mathbf{p}}(t)\mathbf{k}(u) = 0 \tag{60.4}$$

with $\tilde{\mathbf{p}} = (p_n, p_{n-1}, \ldots, p_1)$ and the initial conditions in the form

$$\mathbf{k}(u(t_0)) = \mathbf{b}. \tag{60.5}$$

We note that the equation

$$v = \mathbf{L}u \tag{60.6}$$

defines a linear mapping $\mathbf{L}$: $C^n[a, b] \to C[a, b]$.

The solutions of (60.1) constitute the kernel of the mapping; it is easily verified that the solutions form a vector subspace of $C^n[a, b]$ which we call the *solution space* of equation (60.1).

The properties of this solution space can be inferred from the corresponding properties of the *companion equation*

$$\dot{\mathbf{y}} = \mathbf{A}(t)\mathbf{y} \tag{60.7}$$

where

$$\mathbf{A}(t) = \begin{bmatrix} 0 & 1 & 0 & \dots & 0 \\ 0 & 0 & 1 & \dots & 0 \\ \vdots & \vdots & \vdots & & \vdots \\ -p_n & -p_{n-1} & -p_{n-2} & \cdots & -p_1 \end{bmatrix} \tag{60.8}$$

In scalar form this equation becomes

$$\begin{aligned} \dot{y}_1 &= y_2 \\ \dot{y}_2 &= y_3 \\ \vdots \quad & \quad \vdots \\ \dot{y}_{n-1} &= y_n \\ \dot{y}_n &= -p_n y_1 - p_{n-1} y_2 - p_{n-2} y_3 - \cdots - p_1 y_n \end{aligned} \tag{60.9}$$

the last equation being merely

$$y_1^{(n)} = -\tilde{\mathbf{p}}\mathbf{k}(y_1) \tag{60.10}$$

which is equivalent to

$$\mathbf{L}y_1 = 0.$$

Conversely, the Wronksian vector of any solution of $\mathbf{L}u = 0$ is readily seen to have components which satisfy (60.9).

Theorem 60.1 *The solution space of the nth-order homogeneous equation*

$$u^{(n)} + \sum_{j=1}^{n-1} p_j u^{(n-j)} = 0$$

has dimension n.

Proof. Let $\mathbf{u}$ represent the vector whose components are the components of the first row of some specific fundamental matrix for (60.7). Since the matrix must be a Wronksian matrix it can be represented by $\mathbf{K}(\mathbf{u})$. The components of $\mathbf{u}$ obviously belong to $C^n[a, b]$, and each one must be a solution of (60.1).

Appendices

§A.1 The Laplace transform

The *Laplace transform* of an integrable function $f(x)$ on the positive real line is the function $\bar{f}(p)=\mathscr{L}[f(x);p]$ defined by the equation

$$\bar{f}(p)=\int_0^\infty e^{-px}f(x)\,dx. \tag{A1.1}$$

It is well defined for all $p\in\mathbb{C}$ such that $\operatorname{Re} p>c$ provided $|f(x)|<K\,e^{cx}$ for some finite K. The least value of $\operatorname{Re} p$ such that (A1.1) is defined for all $\operatorname{Re} p$ greater than this value is called the *abscissa of convergence.*

We also make use of the idea of an inverse transform, i.e.

$$\bar{f}(p)=\mathscr{L}[f(x);p]\Rightarrow f(x)=\mathscr{L}^{-1}[\bar{f}(p);x] \tag{A1.2}$$

For instance

$$\mathscr{L}[e^{ax};p]=(p-a)^{-1}\qquad(\operatorname{Re} p>a), \tag{A1.3}$$

$$\mathscr{L}[x^{\nu};p]=p^{-\nu-1}\Gamma(\nu+1)\qquad(\operatorname{Re} p>0,\ \nu>-1). \tag{A1.4}$$

Also, since

$$\int_0^\infty e^{-px}\cos\omega x\,dx=\frac{p}{\omega}\int_0^\infty e^{-px}\sin\omega x\,dx=$$

$$=\frac{p}{\omega}\left\{\frac{1}{\omega}-\frac{p}{\omega}\int_0^\infty e^{-px}\cos\omega x\,dx\right\},$$

it follows that

$$\mathscr{L}[\cos\omega x;p]=p(p^2+\omega^2)^{-1},\qquad \mathscr{L}[\sin\omega x;p]=\omega(p^2+\omega^2)^{-1}. \tag{A1.5}$$

Two rules which are useful in the calculation of Laplace transforms are

$$\mathscr{L}[f(ax);p]=a^{-1}\mathscr{L}[f(x);p/a]\qquad(a>0), \tag{A1.6}$$

$$\mathscr{L}[e^{ax}f(x);p]=\mathscr{L}[f(x);p-a]. \tag{A1.7}$$

To apply the Laplace transform to the solution of differential equations we need a formula by means of which to calculate the Laplace transform of the derivative of a function in terms of the Laplace transform of the function itself. We have

$$\mathscr{L}[f'(x); p] = \int_0^\infty e^{-px} f'(x)\,dx =$$

$$= [f(x)\,e^{-px}]_0^\infty + p\int_0^\infty f(x)\,e^{-px}\,dx.$$

Now for the Laplace transform of $f(x)$ to exist it is necessary that $e^{-px} f(x) \to 0$ as $x \to \infty$, so that we have the result

$$\mathscr{L}[f'(x); p] = p\bar{f}(p) - f(0). \qquad \text{(A1.8)}$$

From this we readily deduce that

$$\mathscr{L}[f''(x); p] = p\mathscr{L}[f'(x); p] - f'(0)$$

and hence that

$$\mathscr{L}[f''(x); p] = p^2\bar{f}(p) - pf(0) - f'(0). \qquad \text{(A1.9)}$$

Proceeding in this way we can easily show that

$$\mathscr{L}[f^{(m)}(x); p] = p^m\bar{f}(p) - \sum_{r=0}^{m-1} p^{m-r-1} f^{(r)}(0). \qquad \text{(A1.10)}$$

In particular, if $f^{(r)}(0) = 0$, $r = 0, 1, \ldots, m-1$ we find that

$$\mathscr{L}[f^{(m)}(x); p] = p^m\bar{f}(p). \qquad \text{(A1.11)}$$

It should also be noted that

$$\mathscr{L}[xf(x); p] = -\frac{d\bar{f}}{dp}, \qquad \text{(A1.12)}$$

a result which is sometimes used in its inverse form

$$\mathscr{L}^{-1}\left[\frac{d\bar{f}}{dp}; x\right] = -xf(x). \qquad \text{(A1.13)}$$

If we introduce the convolution

$$(f * g)(x) = \int_0^x f(t)g(x-t)\,dt \qquad (x > 0), \qquad \text{(A1.14)}$$

we see that

$$\mathscr{L}[g * f; p] = \int_0^\infty e^{-px}\, dx \int_0^x f(t) g(x-t)\, dt =$$

$$= \int_0^\infty f(t)\, dt \int_t^\infty g(x-t)\, e^{-px}\, dx =$$

$$= \int_0^\infty f(t)\, dt\, e^{-pt} \int_0^\infty g(u)\, e^{-pu}\, du$$

i.e.

$$\mathscr{L}[f * g; p] = \bar{f}(p)\bar{g}(p) \tag{A1.15}$$

or

$$\mathscr{L}^{-1}[\bar{f}(p)\bar{g}(p); x] = \int_0^x f(t) g(x-t)\, dt. \tag{A1.16}$$

If we combine equations (A1.15) and (A1.8) we obtain the result

$$\mathscr{L}^{-1}[p f(p) g(p); x] = \frac{d}{dx} \int_0^x f(t) g(x-t)\, dt \tag{A1.17}$$

The result expressed by equation (A1.15) is called the convolution theorem for the Laplace transform.

If we suppose that $f(x)$ can be expanded in a Maclaurin series

$$f(x) = \sum_{r=0}^{\infty} \frac{f^{(r)}(0)}{r!} x^r$$

for all $x > 0$, then

$$\bar{f}(p) = \sum_{r=0}^{\infty} \frac{f^{(r)}(0)}{p^{r+1}},$$

from which it follows that

$$\lim_{p \to 0} p\bar{f}(p) = f(0) \tag{A1.18}$$

It can also be shown that

$$\lim_{p \to 0} p\bar{f}(p) = \lim_{x \to \infty} f(x). \tag{A1.19}$$

Table of Laplace Transforms

	$f(t)$	$\bar{f}(p)$
1	1	p^{-1}
2	$t^n \quad (n=1,2,3,\ldots)$	$p^{-n-1}n!$
3	e^{at}	$(p-a)^{-1}, \quad p>a$
4	$\cosh(at)$	$p(p^2-a^2)^{-1}, \quad p>a$
5	$\sinh(at)$	$a(p^2-a^2)^{-1}, \quad p>a$
6	$t\cosh(at)$	$(p^2+a^2)(p^2-a^2)^{-2}, \quad p>a$
7	$t\sinh(at)$	$2ap(p^2-a^2)^{-2}, \quad p>a$
8	$\frac{1}{2}(at)\cosh(at)-\frac{1}{2}\sinh(at)$	$a^3(p^2-a^2)^{-2}, \quad p>a$
9	$\cosh(at)+\frac{1}{2}(at)\sinh(at)$	$p^3(p^2-a^2)^{-2}, \quad p>a$
10	$\cos(at)$	$p(p^2+a^2)^{-1}$
11	$\sin(at)$	$a(p^2+a^2)^{-1}$
12	$t\cos(at)$	$(p^2-a^2)(p^2+a^2)^{-2}$
13	$t\sin(at)$	$2ap(p^2+a^2)^{-2}$
14	$\frac{1}{2}\sin(at)-\frac{1}{2}(at)\cos(at)$	$a^3(p^2+a^2)^{-2}$
15	$\frac{1}{2}\sin(at)+\frac{1}{2}(at)\cos(at)$	$ap^2(p^2+a^2)^{-2}$
16	$\cos(at)-\frac{1}{2}(at)\sin(at)$	$p^3(p^2+a^2)^{-2}$
17	$J_0(at) \quad (a>0)$	$(p^2+a^2)^{-1/2}$
18	$tJ_0(at)$	$p(p^2+a^2)^{-3/2}$
19	$tJ_1(at)$	$a(p^2+a^2)^{-3/2}$
20	$t^{n+1}J_n(at)$	$\dfrac{(2n+1)!}{2^n n!}\cdot\dfrac{a^n p}{(a^2+p^2)^{n+3/2}}$
21	$t^n J_n(at)$	$\dfrac{(2n)!}{2^n n!}\cdot\dfrac{a^n}{(a^2+p^2)^{n+1/2}}$

§A.2 $e^{t\mathbf{A}}$ where $\mathbf{A}$ is an $n\times n$ matrix

Definition If $\mathbf{A}$ is an $n\times n$ matrix with constant elements we define $e^{\mathbf{A}t}$ to be $\mathbf{\Omega}(t,0)$, the principal matrix solution of $\dot{\mathbf{x}}=\mathbf{A}\mathbf{x}$, i.e. $\mathbf{\Omega}(0,0)=\mathbf{I}$.

The notation is suggested by the fact that the solution of the scalar equation

$$\dot{x}(t)=ax(t), \qquad x(0)=1$$

is e^{at}.

Theorem A2.1 $e^{\mathbf{A}t}$ *has the following properties.*

(i) *The general solution of* $\dot{\mathbf{x}}=\mathbf{A}\mathbf{x}$ *is* $\mathbf{x}(t)=e^{\mathbf{A}t}\mathbf{c}$ *where* $\mathbf{c}$ *is a constant vector;*
(ii) $e^{\mathbf{A}(t+s)}=e^{\mathbf{A}t}e^{\mathbf{A}s}$;
(iii) $(e^{\mathbf{A}t})^{-1}=e^{-\mathbf{A}t}$;
(iv) *If* $\mathbf{Y}(t)$ *is a fundamental matrix of* $\dot{\mathbf{x}}=\mathbf{A}\mathbf{x}$, *then* $e^{\mathbf{A}t}=\mathbf{Y}(t)\mathbf{Y}^{-1}(0)$;
(v) $e^{\mathbf{A}t}=\mathbf{I}+\sum_{j=1}^{\infty}\frac{t^j}{j!}\mathbf{A}^j$.

Proof. (i) It follows immediately from the definition of $e^{\mathbf{A}t}$ that the initial value problem

$$\dot{\mathbf{x}}=\mathbf{A}\mathbf{x} \qquad \mathbf{x}(0)=\mathbf{x}^0$$

has solution $\mathbf{x}(t)=e^{\mathbf{A}t}\mathbf{x}^0$. Taking $\mathbf{x}^0=\mathbf{c}$ we get the stated result.
(ii) Both $e^{\mathbf{A}(t+s)}$ and $e^{\mathbf{A}t}e^{\mathbf{A}s}$ are solutions of the initial value problem $\dot{\mathbf{x}}=\mathbf{A}\mathbf{x}$, $\mathbf{x}(0)=e^{\mathbf{A}s}$ and the result follows from the uniqueness theorem.
(iii) Putting $s=-t$ we have $e^{\mathbf{A}t}e^{-\mathbf{A}t}=e^{\mathbf{A}0}=\mathbf{I}$ implying that $(e^{\mathbf{A}t})^{-1}=e^{-\mathbf{A}t}$.
(iv) Follows directly from the definition.
(v) $\mathbf{\Omega}(t)=\mathbf{\Omega}(t,0)$ satisfies the integral equation

$$\mathbf{\Omega}(t)=\mathbf{I}+\int_0^t \mathbf{A}\mathbf{\Omega}(s)\,ds.$$

We attempt to solve this equation by the method of successive approximations according to the scheme

$$\mathbf{\Omega}_{k+1}(t)=\mathbf{I}+\int_0^t \mathbf{A}\mathbf{\Omega}_k(s)\,ds, \qquad \mathbf{\Omega}_0(t)=\mathbf{I}$$

From this we prove, by induction, that

$$\mathbf{\Omega}_k(t)=\mathbf{I}+\mathbf{A}t+\frac{1}{2!}\mathbf{A}^2t^2+\cdots+\frac{1}{k!}\mathbf{A}^kt^k.$$

It can be shown that the series on the right converges uniformly on any finite interval so that (*v*) follows.

In spite of the apparent simplicity, the matrix $e^{\mathbf{A}t}$ is a rather complicated entity and caution must be exercised in dealing with it. For example

$$\mathbf{B}\,\mathrm{e}^{\mathbf{A}t}=\mathrm{e}^{\mathbf{A}t}\,\mathbf{B}$$

if and only if

$$\mathbf{AB}=\mathbf{BA}.$$

We do not know in what precise sense $\mathrm{e}^{\mathbf{A}t}$ behaves like its scalar counterparts e^{at}, nor do we have a simple means of computing it. We find that to compute $\mathrm{e}^{\mathbf{A}t}$ we must investigate the eigenvalues and eigenvectors of **A**.

Lemma $\mathbf{x}(t)=\mathrm{e}^{\lambda t}\,\mathbf{v}$ *is a non-null solution of* $\dot{\mathbf{x}}(t)=\mathbf{A}\mathbf{x}(t)$ *if and only if* λ *is an eigenvalue of* **A** *and* **v** *is a corresponding eigenvector.*

Proof. If $\mathbf{x}(t)=\mathrm{e}^{\lambda t}\,\mathbf{v}$, then $\dot{\mathbf{x}}=\lambda\,\mathrm{e}^{\lambda t}\,\mathbf{v}$ and this is a solution if and only if

$$\lambda\,\mathrm{e}^{\lambda t}\mathbf{v}=\mathbf{A}\,\mathrm{e}^{\lambda t}\mathbf{v},$$

i.e. if and only if

$$\mathbf{A}\mathbf{v}=\lambda\mathbf{v},$$

which is the stated result.

Theorem A.2 *Suppose that* **A** *has eigenvalues* $\lambda_1,\ldots,\lambda_n$ *(not necessarily distinct) and that there are corresponding eigenvectors* $\mathbf{v}_1,\ldots,\mathbf{v}_n$, *which are* **linearly independent**. *Then*

$$\mathrm{e}^{\mathbf{A}t}=\mathbf{X}(t)\mathbf{X}^{-1}(0)$$

where $\mathbf{X}(t)$ *is the fundamental matrix*

$$\mathbf{X}(t)=(\mathbf{v}_1\,\mathrm{e}^{\lambda_1 t},\ldots,\mathbf{v}_n\,\mathrm{e}^{\lambda_n t}).$$

Proof. The solutions $\{\mathbf{v}_j\,\mathrm{e}^{\lambda_j t}\}_{j=1}^{n}$ are linearly independent since their Wronskian $W(t)$ satisfies

$$W(0)=\det(\mathbf{v}_1,\ldots,\mathbf{v}_n)\neq 0.$$

Then $\mathbf{X}(t)$ is a fundamental matrix and

$$\mathrm{e}^{\mathbf{A}t}=\mathbf{X}(t)\mathbf{X}^{-1}(0).$$

There exist linearly independent eigenvectors in the following two cases:

(1) The eigenvalues are distinct
(2) The matrix A is symmetric.

Of course, eigenvalues and eigenvectors may be complex. However, we always obtain real solutions by equating real and imaginary parts of complex solutions.

Example A2.1 Let

$$\mathbf{A}=\begin{bmatrix}1 & -1 & 4\\ 3 & 2 & -1\\ 2 & 1 & -1\end{bmatrix} \quad\text{and}\quad \det(\mathbf{A}-\lambda\mathbf{I})=\begin{vmatrix}1-\lambda & -1 & 4\\ 3 & 2-\lambda & -1\\ 2 & 1 & -1-\lambda\end{vmatrix}$$

Then

$$\det(\mathbf{A}-\lambda\mathbf{I})=0 \Rightarrow (\lambda-1)(\lambda-3)(\lambda+2)=0.$$

The eigenvalues are $\lambda_1=-2$, $\lambda_2=1$, $\lambda_3=3$ and the corresponding eigenvectors are

$$\mathbf{v}_1=\begin{bmatrix}-1\\ 1\\ 1\end{bmatrix},\quad \mathbf{v}_2=\begin{bmatrix}-1\\ 4\\ 1\end{bmatrix},\quad \mathbf{v}_3=\begin{bmatrix}1\\ 2\\ 1\end{bmatrix},$$

so

$$\mathbf{X}(t)=\begin{bmatrix}-e^{-2t} & -e^{t} & e^{3t}\\ e^{-2t} & 4e^{t} & 2e^{3t}\\ e^{-2t} & e^{t} & e^{3t}\end{bmatrix}$$

$$\mathbf{X}(0)=\begin{bmatrix}-1 & -1 & 1\\ 1 & 4 & 2\\ 1 & 1 & 1\end{bmatrix} \Rightarrow \mathbf{X}^{-1}(0)=\begin{bmatrix}-\frac{1}{3} & -\frac{1}{3} & 1\\ -\frac{1}{6} & \frac{1}{3} & -\frac{1}{2}\\ \frac{1}{2} & 0 & \frac{1}{2}\end{bmatrix}$$

so

$$e^{\mathbf{A}t}=\begin{bmatrix}-e^{-2t} & -e^{t} & e^{3t}\\ e^{-2t} & 4e^{t} & 2e^{3t}\\ e^{-2t} & e^{t} & e^{3t}\end{bmatrix}\begin{bmatrix}-\frac{1}{3} & -\frac{1}{3} & 1\\ -\frac{1}{6} & \frac{1}{3} & -\frac{1}{2}\\ \frac{1}{2} & 0 & \frac{1}{2}\end{bmatrix}$$

$$=\begin{bmatrix}\frac{1}{2}e^{3t}+\frac{1}{6}e^{t}+\frac{1}{3}e^{-2t} & \frac{1}{3}(e^{-2t}-e^{t}) & \frac{1}{2}e^{3t}+\frac{1}{2}e^{t}-e^{-2t}\\ e^{3t}-\frac{2}{3}e^{t}-\frac{1}{3}e^{-2t} & \frac{4}{3}e^{t}-\frac{1}{3}e^{-2t} & e^{3t}-2e^{t}+e^{-2t}\\ \frac{1}{2}e^{3t}-\frac{1}{6}e^{t}-\frac{1}{3}e^{-2t} & \frac{1}{3}(e^{t}-e^{-2t}) & \frac{1}{2}e^{3t}-\frac{1}{2}e^{t}+e^{-2t}\end{bmatrix}$$

Example A2.2 Let

$$\mathbf{A}=\begin{bmatrix} 1 & -3 & 1 \\ 0 & -2 & 1 \\ 0 & 0 & -1 \end{bmatrix}, \det(\mathbf{A}-\lambda\mathbf{I})=(1-\lambda)(-2-\lambda)(-1-\lambda).$$

Then the eigenvalues are $\lambda_1=-2, \lambda_2=-1, \lambda_3=1$ and the corresponding eigenvalues are

$$\mathbf{v}_1=\begin{bmatrix} 1 \\ 1 \\ 0 \end{bmatrix}, \quad \mathbf{v}_2=\begin{bmatrix} 1 \\ 1 \\ 1 \end{bmatrix}, \quad \mathbf{v}_3=\begin{bmatrix} 1 \\ 0 \\ 0 \end{bmatrix},$$

so we may take

$$\mathbf{X}(t)=\begin{bmatrix} e^{-2t} & e^{-t} & e^{t} \\ e^{-2t} & e^{-t} & 0 \\ 0 & e^{-t} & 0 \end{bmatrix}, \quad \mathbf{X}^{-1}(0)=\begin{bmatrix} 0 & 1 & -1 \\ 0 & 0 & 1 \\ 1 & -1 & 0 \end{bmatrix}.$$

Hence

$$e^{\mathbf{A}t}=\begin{bmatrix} e^{t} & e^{-2t}-e^{t} & e^{-t}-e^{-2t} \\ 0 & e^{-2t} & e^{-t}-e^{-2t} \\ 0 & 0 & e^{-t} \end{bmatrix}$$

The procedure has to be modified if the eigenvectors corresponding to a multiple eigenvalue are not linearly independent.

If λ is a multiple eigenvalue (of multiplicity n) which has only one eigenvector $\mathbf{v}^{(1)}$, then we calculate $\mathbf{v}^{(r)}$, $(r=2,\ldots,n)$ where $\mathbf{v}^{(r)}$ is the eigenvector satisfying

$$(\mathbf{A}-\lambda\mathbf{I})^{r-1}\mathbf{v}^{(r)}=0.$$

We then take as the relevant columns of $\mathbf{X}(t)$ the column vectors

$$\mathbf{x}^{(1)}(t)=e^{\lambda t}\,\mathbf{v}^{(1)}(t), \mathbf{x}^{(r)}(t)=$$

$$=e^{\lambda t}\left\{\mathbf{I}+\sum_{s=2}^{r} t^{s-1}(\mathbf{A}-\lambda\mathbf{I})^{s-1}\right\}\mathbf{v}^{(s)} \qquad (r=2,\ldots,n)$$

Example A2.3 Let

$$\mathbf{A}=\begin{bmatrix} 0 & 5 & -3 \\ 1 & 0 & 1 \\ 2 & -4 & 4 \end{bmatrix}, \quad \det(A-\lambda I)=(1-\lambda)^2(2-\lambda).$$

Then the eigenvalues are $\lambda_1=1, \lambda_2=1, \lambda_3=2$, $\mathbf{v}^{(1)}$ satisfying $(\mathbf{A}-\mathbf{I})\mathbf{v}^{(1)}=0$ can be taken to be $[-1, 1, 2]^T$ and $\mathbf{v}^{(2)}$ satisfying $(\mathbf{A}-\mathbf{I})\mathbf{v}^{(2)}=0$ can be taken to be $[0, 1, 2]^T$. Also $\mathbf{v}^{(3)}$ the eigenvector of λ_3 can be taken as $[1, 1, 1]^T$. Hence we may assume the columns of $\mathbf{X}(t)$ to be

$$\mathbf{x}^{(1)}(t)=\begin{bmatrix}-e^t\\ e^t\\ 2e^t\end{bmatrix}, \quad \mathbf{x}^{(2)}(t)=e^t\{\mathbf{I}+t(\mathbf{A}-\mathbf{I})\}\mathbf{v}^{(2)}=$$

$$=\begin{bmatrix}-t\,e^t\\ (1+t)\,e^t\\ 2(1+t)\,e^t\end{bmatrix}, \quad \mathbf{x}^{(3)}(t)=\begin{bmatrix}e^{2t}\\ e^{2t}\\ e^{2t}\end{bmatrix}$$

from which we deduce that

$$\mathbf{X}(t)=\begin{bmatrix}-e^t & -t\,e^t & e^{2t}\\ e^t & (1+t)\,e^t & e^{2t}\\ 2\,e^t & 2(1+t)\,e^t & e^{2t}\end{bmatrix}, \quad \mathbf{X}(0)=\begin{bmatrix}-1 & 0 & 1\\ 1 & 1 & 1\\ 2 & 2 & 1\end{bmatrix},$$

$$\mathbf{X}^{-1}(0)=\begin{bmatrix}-1 & 2 & -1\\ 1 & -3 & 2\\ 0 & 2 & -1\end{bmatrix}$$

and hence that

$$e^{At}=\mathbf{X}(t)X^{-1}(0)=\begin{bmatrix}(1-t)\,e^t & 2\,e^{2t}-(2+3t)\,e^t & (1-2t)\,e^t-e^{2t}\\ t\,e^t & 2\,e^{2t}-(1+3t)\,e^t & (1+2t)\,e^t-e^{2t}\\ 2t\,e^t & 2\,e^{2t}-2(1+3t)\,e^t & 2(1+2t)\,e^t-e^{2t}\end{bmatrix}$$

Problems

Derive the general solutions of equations 1–36:

1 $(x^2-x)y' = y^2 + y$
2 $xy' + y^2 = 1$
3 $(1+y)y' = x^2(1-y)$
4 $xy' + (2x^2-1)\cot y = 0$
5 $2xy(x+1)y' = y^2 + 1$
6 $x\sqrt{(y^2-1)} + yy'\sqrt{(x^2-1)} = 0$
7 $\sqrt{(1-x^2)}y' = 1 + y^2$
8 $y' + (1-y^2)\tan x = 0$
9 $x(1-x^2)y' = (x^2-x+1)y$
10 $xy^3y' = 1 - x^2 + y^2 - x^2y^2$
11 $(x^2+a^2)y' = (y+b)\{x + \sqrt{(x^2+a^2)}\}$
12 $x^2(y+a)^2(y'-1) = y^2 - 2ax^2y + a^2$
13 $(x^2+y^2)y' = xy$
14 $(x^2-2xy-y^2)y' = x^2 + 2xy - y^2$
15 $(x^2+2xy)y' = y^2 - 2xy$
16 $x(x^2-6y^2)y' = 4y(x^2+3y^2)$
17 $y - xy' = \sqrt{(x^2+y^2)}$
18 $x^2y = (x^3+ay^3)y'$
19 $x(x-ay)y' = y(y-ax)$
20 $(x^2+xy+ay^2)y' = ax^2 + xy + y^2$
21 $x(x+y)y' = x^2 + y^2$
22 $x(x^2+axy+y^2)y' = (x^2+bxy+y^2)y$
23 $(x+y)^2y' = x^2 - 2xy + 5y^2$
24 $xy' = y - x\cos^2(y/x)$
25 $(3y-x)y' = 3x - y + 4$
26 $(5x-y+1)y' + x - 5y + 5 = 0$
27 $(4y+x)y' = y - 4x$
28 $(9x+2y+19)y' = 2x + 6y - 18$
29 $(9x+21y+3)y' = 7x - 5y + 45$
30 $(7x-16y+140)y' + 8x + y + 25 = 0$
31 $(2x-4y+5)y' = x - 2y + 3$
32 $(y+ax+b)y' = (y+ax-b)$

33 $x^2y' = (2x - y + 1)^2$
34 $(x + y + a + b)^2 y' = 2(y + a)^2$
35 $(x - y)^2 y' = (x - y + 1)^2$
36 $(x + y)^2 y' = (x + y + 2)^2$

Show that each of the equations 37–44 are exact and integrate them:

37 $(hx + by)y' + ax + hy = 0$
38 $(2x^2 + 3y^2)y' + 3x^2 + 4xy = 0$
39 $\{(a + 2h)x^2 + 2(b + 2h)xy + 3by^2\}y' + 3ax^2 + 2(a + 2h)xy + (b + 2h)y^2 = 0$
40 $(4x^3y - 12x^2y^2 + 5x^2 + 3x)y' + 6x^2y^2 - 8xy^3 + 10xy + 3y = 0$
41 $x^2y'/y + 2x \log |y| = 0$
42 $(\cos x - x \cos y)y' - \sin y - y \sin x = 0$
43 $\left(\dfrac{x + a}{y + b}\right)^2 y' = 2\dfrac{x + a}{y + b}$
44 $\dfrac{(1 - x^2)y' + (1 - y^2)}{(1 + xy)^2} = 0$

Show that each of the equations 45–66 has an integrating factor of one or other of the types $\mu(x)$, $\mu(y)$, $\mu(x + y)$, $\mu(xy)$ and hence find their general solutions:

45 $y' + x + y = 0$
46 $8xy^3y' + x^3 + y^4 = 0$
47 $(x^2y^2 - 1)y' + xy^3 = 0$
48 $(1 - x^2)y' = xy - 1$
49 $(1 + xy)y' + y^2 = 0$
50 $\{1 + (x + y) \tan y\}y' + 1 = 0$
51 $3x(x + y^2)y' = 2y^3 + 3xy - x^3$
52 $(y^2 - x^2 + 1)y' = x^2 - y^2 + 1$
53 $(x^2 + a^2)(x + 2y)y' + (2x^2 + xy + a^2)y = 0$
54 $(5x^6y^7 + x^3)y' = y^3 - 3x^5y^8$
55 $(xy + y^2 + y + 1)y' + y + 1 = 0$
56 $(4x^3 + x + 5y)y' + (7x^3 + 3x^2y + 4y) = 0$

Find the general solution of each of the equations 57–104:

57 $xy' - y = x^3 + 1$
58 $xy' + y = x \log |x|$
59 $(x + 1)y' - y = 3x^4 + 4x^3$
60 $(x^2 + a)y' = xy + a$
61 $(1 - x^2)y' = xy + 1$
62 $(1 - x^2)y' = 1 - x^2 + y$
63 $(1 + x^2)y' + xy = 3x^3$

64 $(2xy'+y)\sqrt{(1+x)}=1+2x$
65 $y'\sqrt{(1+x^2)}+y=2x$
66 $2x(x-1)y'+(2x-1)y=x$
67 $x^2y'+xy=x^2+x+1$
68 $2(x^2+x+1)y'+(2x+1)y=8x^2+1$
69 $(x^2-1)y'-xy=x^2$
70 $x(1-x^2)y'+(2x^2-1)y=x^3-x^5$
71 $(1+x^2)y'+2xy=\tan x$
72 $2(1-x^2)y'-(1+x)y=\sqrt{(1-x^2)}$
73 $2(1-x)y'-y=4x\sqrt{(1-x)}$
74 $(1+x^2)^2y'+(1+x)(1+x^2)y=2x$
75 $xy'+2y=\sin x$
76 $y'\cos x+y\sin x=\cos^2 x$
77 $y'+y\tan x=\sin 2x$
78 $y'\sin x\cos^3 x+y\cos 2x\cos^2 x=1$
79 $y'+y\cos x=\frac{1}{2}\sin 2x$
80 $y'\sin x-y\cos x=e^x\sin^2 x$
81 $y'-y=x+\sin x$
82 $\frac{1}{2}y'=y\tan 2x+1+\sec 2x$
83 $y'+2y=x^2+3\cosh x$
84 $y'\cos x+y\sin x=1+\tan x$
85 $y'+y\tanh x=6\,e^{2x}$
86 $y'\sin^2 x-y\tan x=\tan x-\tan^3 x$
87 $y'\cos x-3y\sin x=\cot x$
88 $y'+2y\operatorname{cosec} 2x=2\cot^2 x\cos 2x$
89 $y'\tan x+y=\sin 3x+\operatorname{cosec} 2x$
90 $y'\cot x-y=\operatorname{cosec} 2x+\cos 2x$
91 $(1-x^2)y'+xy=x\arcsin x+(1-x)\sqrt{(1-x^2)}$
92 $x(x+1)y'-(x^2+x-1)y=(x+1)(x^2-1)$
93 $xy'=4y-4\sqrt{y}$
94 $2xy'-y=(2x^3-1)/y$
95 $xy'=y+2xy^2$
96 $(1+x^2)y'+xy=xy^2$
97 $x^3y'=2x^2y+y^3$
98 $xy'+y=y^2\log|x|$
99 $xy'+y=xy^3$
100 $x^2y'+xy+\sqrt{y}=0$
101 $3xy'-2y=xy^4$
102 $y'\cos x+y\,\mathrm{sim}\,x+y^3=0$
103 $y'+2y=2xy^{3/2}$
104 $2(1+x)yy'+2x-3x^2+y^2=0$

105 Show that the equation

$$y' = (y-1)(xy-y-x)$$

has particular solution $y=1$ and derive the general solution.

106. Verify that $y=(x+1)/x^2$ is a particular solution of

$$y' + y^2 = x^{-4}$$

and derive the general solution.

Show that $y=x$ is a particular solution of each of the equations 107–110 and complete the integration:

107 $xy' + (m-1)ax^m(y-x)^2 + y - 2x = 0$

108 $y' + xy^2 - (2x^2+1)y + x^3 + x - 1 = 0$

109 $(x^2+a)y' + 2y^2 - 3xy - a = 0$

110 $2x^2y' = 2xy + (y^2 - x^2)(x \cot x - 1)$

Integrate equations 111–118 by making a change of variable:

111 $x(y'+1) + \tan(x+y) = 0$

112 $xy' - y = x^2 + y^2$

113 $(1-x^2)yy' + xy^2 + 2x^2 = 0$

114 $2(1+x)yy' + y^2 + 2x - 3x^2 = 0$

115 $(x^2+xy+1)y' = (y^2+xy+1)$

116 $(x^2+y^2-a)yy' + x(x^2+y^2+a) = 0$

117 $\{(x^2+y^2-a)x - y\}y' = (x^2+y^2-a)y + x$

118 $(x^2+y^2-a)(x+yy') = 2xy(y-xy')$

119 By changing the dependent variable to v where $v=x^2/y^2$, show that

$$\frac{1}{2}\frac{dv}{dy} = \frac{x}{y^2y'} - \frac{x^2}{y^3};$$

hence, find the general solution of the equation

$$(x^2 - y^4)y' = xy.$$

120 By changing the dependent variable to $v = y/x^2$, find the general solution of the equation

$$(xy' - 2y)^2 = x^2(x^4 - y^2).$$

121 Find the general solution of the equation

$$(x^4-1)(xy'-y) = 2x(y^2 - x^2)$$

by changing the dependent variable from y to v, where $y = vx$.

122 Transform the equation

$$x^m(xy'-y)=y^2-x^2$$

to one in z where

$$y=x\frac{1+z}{1-z}$$

and hence find its general solution when (i) $m=0$, (ii) $m=1$, (iii) $m=2$.

123 Find the equation of the curve whose radius vector equals its tangent.

124 Find the equation of the curve whose normal is bisected by the y-axis.

125 Find the equation of the curve which passes through the point (O, a), $(a>0)$, and whose tangent is of constant length a.

126 Find the equation of the curve such that the area of the trapezium bounded by its tangent, its ordinate and the coordinate axis is a constant.

127 Find the curves with the property that the projection of the ordinate on the normal is constant.

128 Find the curves which have their subtangent equal to the abscissa of the projection of the origin on the tangent.

129 Determine the curves the midpoints of whose normals lie on the parabola $y^2=4ax$.

130 (Cf. Fig. 12 on p.46.) The ordinate QA of a point A on a curve Γ meets the line $y=ax+b$ in the point R; the point S is taken on QA such that $QR=AS$. Determine the equation of Γ if OS is parallel to the tangent at A, for all points A on Γ.

131 Determine the curves for which the sum of lengths of the ordinate and the normal is $2a$, where a is a positive constant.

132 Find the curve whose arc length measured from the point $(a, 0)$ is x^2/a.

If the line through the origin O perpendicular to the line OA meets the tangent at A to a curve Γ in U and the normal in H, OU is called the *polar subtangent*, HO the *polar subnormal*, AU the *polar length* and AH the *polar normal.* Find the curves Γ for which

133 $HO=k^2\cdot OU$;

134 $OU/HO=y^2/x^2$;

135 $AH=a$;

136 H lies on the line $x=a$.

137 Find the equation (in plane polar coordinates r, θ) whose arc length $s = \sqrt{(r^2 + 2ar)}$.

138 Find the curves for which $\phi = m\theta$, where ϕ is defined as the angle QTA in Fig. 12 on p.46.

Find the orthogonal trajectories of the families of curves described by the following equations (in which c is a parameter and a, b are constants):

139 $x^3 - 3xy^2 = c$

140 $y^2 = cx$

141 $x^2 + cy^2 = a,$

142 $x^2 - y^2 + 2cxy = a^2$

143 $x(x^2 + y^2) = c(y^2 - x^2)$

144 $(x^2 - a^2)^2 + (y^2 - b^2)^2 = c^2$

145 $xy = c(x-1)^2$

146 $y = c \log |x|$

147 $y = c(x+a)\,e^x$

148 $\tanh^2 x + \tanh^2 (y + c) = 1$

149 Show that the ellipses of the one-parameter family

$$x^2/c + y^2 = a^2/(c-1)$$

are self-orthogonal.

150 Find the curves orthogonal to the set of rectangular hyperbolas through the point $(a, 0)$ with the y-axis as asymptote.

151 Find the curves orthogonal to the set of ellipses with centre O and common major axis, of length $2a$, along the x-axis.

152 Find the set of curves orthogonal to the set of conics circumscribing the rectangle with vertices $(\pm a, \pm b)$.

153 Find the curves orthogonal to the set of parabolas with common focus O and latus rectum $2l$.

Find the orthogonal trajectories of the families of curves described by plane polar coordinates (r, θ) by the following equations:

154 $r = \tan(\theta + c)$

155 $(r^2 + a^2)\sin\theta + cr = 0$

156 $r = c/(1 + 2\cos\theta)$

157 $r = c(1 + 2\cos\theta)$

158 Find the trajectories at angle $\tan^{-1} m$ of the family of hyperbolas $x^2 - y^2 = c$.

159 Find the trajectories at angle $\tan^{-1} m$ of the family of circles $x^2 + y^2 = 2cx$.

160 Find the trajectories at angle ω of the family of curves $r^p = c\cos(p\theta)$.

Integrate equations 161 and 162, and determine the shape of their integral curves in the vicinity of the origin:

161 $y' = (y - x^2)/mx$

162 $y' = (y^2 - x)/y$.

Solve equations 163–166 for y' and find their general integrals:

163 $xy'^2 - 2yy' - x = 0$

164 $(1 - y'^2)\cos x = 2y'\sin x$

165 $(y' - x)^2 = y' + x$

166 $x^2y'^2 - xyy' + y^2 = 0$.

Obtain the singular solutions of $y = px + f(p)$ when $f(p)$ takes the following forms:

167 p^3

168 $p/(p+1)$

169 $2\sqrt{(ap)}$

170 $\sqrt{(p^2 - 1)}$

171 $\frac{2}{3}(p+1)^{3/2}$

172 $\sqrt{\{(a + 2hp + bp^2)/(ab - h^2)\}}$

173 $ap(p^2 + 1)^{-1/2}$

174 $p - p\log|p|$

175 $\sqrt{(1 - p^2)} - p \arccos p$

176 $\sqrt{(p^2 + 1)} - p\log\{p + \sqrt{(p^2 + 1)}\}$

177 Find the general solution and the singular solution of

$$y = 3px + p^{-1}y^{-4}$$

by first changing the independent variable to $v = y^3$.

178 Find the general solution and the singular solution of

$$y = p + e^x/p.$$

179 Transform the equation

$$(1 + p^2)y^2 = f(x + py)$$

by the substitution $z = \frac{1}{2}(x^2 + y^2)$ and hence integrate it.

180 Find the singular solution of $(1 + p^2)y^2 = (x + py)^3$.

181 Find the general solution and the singular solution of the equation

$$x^4p^3 - x^3yp^2 - x^2y^2p + xy^3 = 1,$$

first making the substitutions $y = vx$, $x = t^{-1/2}$.

Find the general solution (and the singular solution, if any) of each of the equations 182–191:

182 $y = 2xp + p^2$

183 $y = x + ap - a\log|p|$

184 $x-y=2ap-ap^2$
185 $3yp^2=(2p^3-1)x$
186 $y=p^2+p^3$
187 $y=yp^2+2px$
188 $y=xp^2+2p$
189 $y=xp^2+p^3$
190 $(p-x)^3+p(p-x)=1$
191 $a^2(p-b)^2+p(y+bx)^2=0.$
192 Show that if

$$x+yp=f(p)$$

then

$$y=\{c+\int f'(\tan t)\,\mathrm{d}(\sec t)\}\cos t.$$

Hence, integrate

$$x+yp=p^3.$$

193 Find the integral curves of the equation

$$r'^2\sin\theta-2rr'\cos\theta-r(r\sin\theta+a)=0$$

where $r'=\mathrm{d}r/\mathrm{d}\theta$ and a is a constant.

Making the substitution $y=tp$ in each of the equations find the solutions of equations 194–197 in the form of parametric equations for x and y:

194 $(y+p)^2=y-p,$
195 $(y+p)^4=3y+p,$
196 $y^3-3yp+p^3=0$
197 $y^4+p^4=yp(y+p).$
198 Making the substitution $y=2x+tp$, solve

$$(p-1)(y-2x)^2=p^2.$$

199 Determine the constants a, m, n such that the differential equation

$$y^2p^2+2xyp+ay^2+2mx+n=0$$

shall have a real singular solution. Integrate completely in this case.

200 Find the integral curves and p-discriminant locus of the differential equation

$$y^2-x^2-2xyp+m^2\{(x+yp)^2-a^2\}=0.$$

Derive the general solutions of equations 201–214:

201 $y''+y'^2=1$

202 $y''+(x-a)y'^3=0$

203 $2yy''=y'^2$

204 $xy^2y''-xyy'^2+(y^2+a)y'=0$

205 $yy''=y'^2-y'^3$

206 $yy''+y'^2+2a^2y^2=0$

207 $(y-1)y''=2y'^2$

208 $yy''+\frac{1}{2}y'^2=y^{-2}$

209 $xyy''+ny'(xy'-y)=0$

210 $x^2yy''=2(xy')^2+axyy'+ay^2$

211 $(2yy''-y'^2)x^2+y^2=0$

212 $x(1-x)y''+2(1-x)y'=1$

213 $3xy'^2y''=1+y'^3$

214 $yy''-y'^2+y'=0$

215 Show that the form of the general solution of the differential equation

$$yy''=(a+1)y'^2+byy'+cy^2 \qquad (*)$$

depends on the nature of the roots of the quadratic equation

$$au^2+bu+c=0.$$

(i) Given that $y=(c_1\,e^{-max}+c_2\,e^{-nax})^{-1/a}$, c_1 and c_2 real constants, $m\neq n$ both real, show that

$$\frac{(y'-ny)^n}{(y'-my)^m}=cy^{(a+1)(n-m)}$$

(ii) Given that $y=e^{mx}(c_1+c_2x)^{-1/a}$, with a, c_1, c_2 and m real constants, show that

$$\log\frac{y'-my}{y^{a+1}}=c+\frac{my}{y'-my}$$

where c is a constant.

Making use of these results integrate the equation (*) in the usual three cases.

216 Show that the equation

$$(y-x)y''+F(y')=0$$

has a first integral of the form

$$(y-x)f(y')=c_1.$$

Integrate in the special case in which

$$F(p)=(1+p)(1+p^2).$$

217 Show that the equation

$$yy''=my'^2+ayy'+by^2+cy^{m+1}$$

may be reduced to linear form by a substitution

$$y=\begin{cases} u^{\alpha} & (m\neq 1) \\ e^{u} & (m=1), \end{cases}$$

for a suitable choice of the constant α.

Using the method of Problem 217, find the general solution of equations 218–221:

218 $yy''=2y'^2-2y^3$

219 $2yy''=6y'^2+y^2-3y^4$

220 $yy''=y'^2+yy'$

221 $3yy''=2y'^2+36y^2$

Find two-parameter families of curves such that, in the notation of Fig. 12 on p.46 the conditions 222–226 are satisfied:

222 $s=OT$

223 $AN^3=k\rho$

224 $\rho\cos^2\phi=a$

225 $\rho=TN$

226 The ordinate through T bisects the radius of curvature.

227 Form the differential equation whose general solution is

$$y=c_1(1+x)^m+c_2(1-x)^m$$

Show that each of the equations 228–230 has a polynomial solution and hence derive the general solution of each:

228 $xy''-(x+3)y'+y=0$

229 $(1+x^2)y''-2y=0$

230 $x(x-1)y''+(2x-1)y'-6y=0$

231 Show that $y=1/(x+1)$ is a particular integral of the equation

$$x(x+1)y''=\{(n-2)x+n\}y'+ny$$

and hence derive the general solution.

232 Show that $y=x/(1-x)$ is a particular integral of the equation

$$x(1-x)^2y''=2y$$

and hence derive the general solution.

Find the general solution of each of the equations 233–244:

233 $y''+3y'+2y=e^x+\sin x$

234 $y''-3y'-4y=10\cos 2x$

235 $y''+2y'+5y=8\sinh x$

236 $y''+6y'+9y=e^{-3x}\cosh x$

237 $y''-5y'+6y=4x^2 e^x$

238 $y''+2y'+3y=e^{-x}\cos x$

239 $y''-4y'+4y=e^{2x}\cos^2 x$

240 $y'''-3y'+2y=x^2 e^x$

241 $y''-2y'\sin\alpha+y=2\cos\alpha\sin x \qquad (\alpha \neq \frac{1}{2}n\pi)$

242 $y''\cos^2\alpha-y'\sin 2\alpha+y=x^2\exp(x\tan\alpha)$

243 $y^{iv}+2y''+y=24x\sin x$

244 $y^{vi}-y^{iv}-y''+y=4e^x$

Find particular integrals of $y''+y=f(x)$ when $f(x)$ has the forms stated in 245–250:

245 $e^x\sin 2x$

246 $e^{2x}\cos x$

247 $e^x(x^2-1)$

248 $8\cos x\cos 2x$

249 $4x\sin x$

250 $x\cos x-x^2\sin x$

Find particular integrals of $y''-2y'+y=f(x)$ when $f(x)$ has the forms stated in 251–256

251 x^3-6x^2

252 $e^x(1+2x+3x^2)$

253 $x^{-2}e^x$

254 $e^x\sin x$

255 $50\cosh x\cos x$

256 $8x^2 e^{3x}$

Solve the initial-value problems 257–261:

257 $y''-2y'+(1+m^2)y=(1+4m^2)\cos mx, \quad y(0)=1,\ y'(0)=0$

258 $y''+3y'+2y=2x\,e^{-x} \qquad y(1)=y'(1)=0$

259 $y''-2ay'+(a^2+m^2)y=2m\,e^{ax}\cos mx,$

$$y(0)=0,\ y'(0)=m$$

260 $y'''-y''-y'+y=8x\,e^{-x},$

$$y(0)=0,\ y'(0)=1,\ y''(0)=0$$

261 $y^{iv}-2y''+y=12x\,e^{-x},$

$$y(0)=y''(0)=\tfrac{1}{2},\ y'(0)=0,\ y'''(0)=-3$$

Find the general solution of each of the equations 262–270:

262 $2x^2y''-xy'+y=x^2$

263 $x^2y'' - 2xy' + 2y = 2x \log|x|$
264 $x^2y'' + 4xy' + 2y = \log|1+x|$
265 $x^2y'' + xy' + y = \log|x|$
266 $x^2y'' - 2xy' + 2y = x^3 \sin x$
267 $2(x+1)^2y'' - (x+1)y' + y = x$
268 $x^{a+1}y'' - (2a-1)x^a y' + a^2x^{a-1}y = 1$
269 $4x^2y'' + y = \sqrt{x}$
270 $x^2y'' - 2nxy' + n(n+1)y = e^x x^{n+2}$
271 Find the general solution of the equation

$$y'' \cos x + y' \sin x = y \cos^3 x$$

by changing the independent variable from x to t where $t = \sin x$.
272 By making the change of variable $x = \sin t$, find the general solution of the equation

$$(1-x^2)y'' - xy' + y = 0.$$

273 Find the general solution of the equation

$$(1+x^2)y'' + xy' = 4y$$

by changing the independent variable from x to t where $x = \sinh t$.
274 Change the independent variable from x to t, where $x = \tan \frac{1}{2}t$ in the equation

$$(1+x^2)^2y'' + 2x(1+x^2)y' + 4y = 0$$

and hence find its general solution.
275 Find a substitution $t = f(x)$ which reduces the equation

$$2p(x)y'' + p'(x)y' + y = 0$$

to one with constant coefficients.
Find the general solution of each of the equations 276–278:
276 $4xy'' + 2y' + y = 0$
277 $y'' \sin 2x + 2y' \cos 2x + 4y \operatorname{cosec} 2x = 0$
278 $y'' + 2 \cot x\, y' - y = 0$
Find the general solutions of equations 279–281 by first reducing them to normal form
279 $x^2y'' - 2nxy' + \{n(n+1) + m^2x\}y = 0$
280 $y'' \sin^2 x - 2y' \tan x + (2 + \sin^2 x)y = 0$
281 $y'' \cos x + 2y' \sin x - y \cos x = 1 + x$

282 Find the general solution of the equation

$$y'' + (e^{2y} + x)y'^3 = 0$$

by taking y as the independent variable.

283 Find the relation between the coefficients $p(x)$, $q(x)$ if the equation $y'' + py' + qy = 0$ has two real solutions (i) whose product is unity, (ii) the sum of whose squares is unity.

Verify the results in the case $p(x) = 1/x$.

284 Solve $xy'' + (\gamma - x)y' - \alpha y = 0$ by definite integrals.

Find particular integrals of equations 285–289 by the method of variation of parameters:

285 $y'' + y = \sec x$

286 $y'' + 4y = 2 \tan x$

287 $y'' + y = \log |\cos x|$

288 $2x^2y'' + 7xy' + 3y = \cos(\sqrt{x})$

289 $y'' + 2ky' + (k^2 + w^2)y = f(x) \qquad (x > 0)$

290 Show that the operator $xD^2 + (1 - x)D - 1$ may be written $(xD + 1)(D - 1)y$, and hence derive the general solution of the equation

$$xy'' + (1 - x)y' - y = e^x.$$

291 Find the general solution of the equation

$$(x + 2)y'' - (2x + 5)y' + 2y = (x + 1)e^x$$

by first factorizing the operator on the left-hand side.

292 By factorizing the differential operator or otherwise, find the general solution of the equation

$$y'' - 2(n - ax^{-1})y' + (n - 2anx^{-1})y = e^{nx}$$

in which n and a are constants.

Examine the cases $a = \pm\frac{1}{2}$:

293 If

$$D^2 + P(x)D + Q(x) \equiv \{D + u(x)\}\{D + v(x)\}$$

show that u and v satisfy respectively the Riccati equations

$$u' - Pu + u^2 = P' - Q$$

$$v' + Pv - v^2 = Q.$$

Using the method of the last problem find the general solution of each of the equations 294 and 295:

294 $y'' + (\cot x - 2)y' - (\operatorname{cosec}^2 x + 2 \cot x)y = e^x,$

295 $y'' - y' \tan x - 2y/(1+\sin x) = 0$.

296 Show that $y = \sec x$ is a solution of the equation $\mathbf{L}y = 0$, where $\mathbf{L}$ denotes the differential operator

$$x\frac{\mathrm{d}^2}{\mathrm{d}x^2} + (1 - 2x\tan x)\frac{\mathrm{d}}{\mathrm{d}x} - (x + \tan x)$$

and deduce the general solution.

Show also that the initial-value problem

$$\mathbf{L}y(x) = f(x), \quad x > 0, \qquad y(0) = y'(0) = 0$$

has solution

$$y(x) = \sec x \int_0^x \cos t \log(x/t) f(t)\,\mathrm{d}t.$$

297 Solve the equation $y'' + 2xy' + x^2y = 0$ by first reducing it to normal form, and deduce the form of the Green's function for the initial-value problem.

298 Prove that the solution of the equation

$$y'' + (a+b)y' + aby = f(x)$$

(a and b constants) satisfying $y(0) = y'(0) = 0$ is

$$y(x) = \frac{1}{b-a}\int_0^x \{\mathrm{e}^{-a(x-t)} - \mathrm{e}^{-b(x-t)}\} f(t)\,\mathrm{d}t.$$

In particular find the solution corresponding to

$$f(x) = \begin{cases} 1 & 0 < x < 1 \\ 0 & \quad x > 1. \end{cases}$$

299 Solve the equation

$$x^2y'' + x(x^2+1)y' + (x^2-1)y = 0$$

(a) by finding an integrating factor of the form x^a;

(b) by finding a particular solution of the form x^c.

300 Given that $y = v(x)$ is a particular solution of the equation $y'' = f(x)y$ $(a < x < b)$, show that the general solution may be written in the form

$$v(x)\{c_1 + c_2u(x)\}$$

where c_1 and c_2 are arbitrary constants and $u'(x) = \{v(x)\}^{-2}$.

Find a polynomial solution of the equation

$$(x^2+1)y''=2y \qquad (x>0)$$

and hence derive the general solution.

301 Given that the differential equation

$$xy''+xa(x)y'+b(x)y=0$$

has two solutions $y=u(x)$ and $y=u(x)\log x$, find $u(x)$ in terms of $a(x)$ in a form involving an indefinite integral. Hence show that

$$b(x)=\tfrac{1}{4}+\tfrac{1}{2}xa'(x)-\tfrac{1}{2}a(x)+\tfrac{1}{4}\{a(x)\}^2.$$

Given that $b(x)\equiv\tfrac{1}{4}$, $a(1)=1$ and $u(1)=1$ find $a(x)$ and $u(x)$.

302 There exists a function $f(x)$ which satisfies the conditions

$$\text{(i)}\quad f'(x)=f\left(\frac{1}{x}\right), \qquad \text{(ii)}\quad f(1)=1.$$

Prove that it satisfies the differential equation

$$x^2y''(x)+y(x)=0.$$

Solve this equation and deduce that

$$f(x)=x^{1/2}\cos(\tfrac{1}{2}\sqrt{3}\log x)+\frac{1}{\sqrt{3}}x^{1/2}\sin(\tfrac{1}{2}\sqrt{3}\log x).$$

303 Given that

$$(x^2+1)^2z'(x)=2-6x^2+4x(x^2+1)z-(x^2+1)^2z^2$$

and $z=y'(x)/y(x)$, find the linear differential equation of the second order satisfied by y. Show that it has two solutions y_1, y_2 such that $y_2=xy_1$ and find y_1, y_2.

304 Show that the equation $y'+y^2+9x^4=0$ can be reduced by the substitution $y=u'(x)/u(x)$ to the equation $u''+9x^4u=0$. Prove that the complete solution of this equation is

$$u=x^{1/2}\{AJ_{1/6}(x^3)+BJ_{-1/6}(x^3)\}$$

and hence that the solution of the original equation is

$$y=\frac{1}{2x}+3x^2\frac{J'_{1/6}(x^3)+CJ'_{-1/6}(x^3)}{J_{1/6}(x^3)+CJ_{-1/6}(x^3)}$$

where C is an arbitrary constant.

305 Without solving the differential equation

$$\frac{\mathrm{d}u}{\mathrm{d}x}\frac{\mathrm{d}^3u}{\mathrm{d}x^3}=m\left(\frac{\mathrm{d}^2u}{\mathrm{d}x^2}\right)^2$$

where m is a constant, show that it has the property that the functional product $(f\circ g)(x)=f\{g(x)\}$ of two solutions $f(x), g(x)$ is always a solution, if and only if $m=\frac{3}{2}$.

Solve the equation when $m=\frac{3}{2}$ and verify the property.

306 Show that the solution of the initial-value problem

$$xy''(x)+y(x)=0, \qquad y(0)=0, \qquad y'(0)=1$$

is

$$y(x)=\mathscr{L}^{-1}[p^{-2}\,\mathrm{e}^{-1/p};\,p\to x]=x^{1/2}J_1(2x^{1/2}).$$

307 The polynomial $L_n(x)$ is the solution of the system

$$xy''+(1+x)y'+ny=0, \qquad y(0)=1$$

where n is a positive integer. Show that

$$L_n(x)=\mathscr{L}^{-1}\{(p-1)^n p^{-n-1};\,x\}$$

and deduce that

(i)
$$L_n(x)=\sum_{r=0}^{n}(-1)^r\binom{n}{r}\frac{x^r}{r!};$$

(ii)
$$\sum_{n=0}^{\infty}L_n(x)\frac{t^n}{n!}=\mathrm{e}^t J_0\{2\sqrt{(xt)}\}.$$

308 Solve the differential equation

$$y''+y=x^{-1}$$

by the method of variation of parameters, and deduce that the function

$$\int_0^\infty \frac{\sin t}{t+x}\,\mathrm{d}t$$

is a particular solution of it for $x>0$. Show also that this function may be defined by the integral

$$\int_0^\infty \frac{\mathrm{e}^{-xu}}{1+u^2}\,\mathrm{d}u.$$

309 Given that $y(x)=\cos(2m\sin^{-1}x)$, show that

$$(1-x^2)y''(x)-xy'(x)+4m^2y(x)=0;$$

deduce that, for all positive integers n

$$y^{(n+2)}(0)+(4m^2-n^2)y^{(n)}(0)=0.$$

Hence show that if m is a positive integer the Maclaurin series for $y(x)$ has a finite number of terms and is the series

$$1+\sum_{k=1}^{m}(-4)^k\frac{m^2(m^2-1^2)\cdots(m^2-(k-1)^2)}{(2k)!}x^{2k}.$$

Check this answer in the cases $m=1$ and $m=2$ by considering the formulae for $\cos 2\theta$ and $\cos 4\theta$ in terms of powers of $\sin\theta$.

310 Given that $y=\frac{1}{2}(\sin^{-1}x)^2$ show that

$$(1-x^2)y''(x)-xy'(x)=1.$$

Deduce that

$$y^{(n+2)}(0)=n^2y^{(n)}(0).$$

Hence obtain the Maclaurin series for y, stating clearly the coefficients of x^{2m} and x^{2m+1}.

311 Verify that, if $y=\{x+\surd(x^2+1)\}^k$, then $(x^2+1)y''+xy'-k^2y=0$. Deduce a relation between $y^{(n+2)}$, $y^{(n+1)}$ and $y^{(n)}$, and find an explicit expression for $y^{(n)}(0)$, distinguishing between the cases n even and n odd.

312 Given that $y(x)$ is any solution of the differential equation

$$(1-x^2)y''(x)-xy'(x)+a^2y(x)=0$$

where a is a real constant, show that

$$y^{(n+2)}(0)=(n^2-a^2)y^{(n)}(0) \qquad (n>0).$$

Hence obtain the Maclaurin expansion of the particular solution satisfying $y(0)=1$, $y'(0)=0$, showing clearly the coefficient of the terms in x^{2n} and x^{2n+1}.

Under what conditions is this solution a polynomial?

313 Given that $y(x)$ is any solution of the differential equation

$$4(1+x^2)y''(x)+4xy'(x)-y(x)=0,$$

prove that, for all integers $n>1$,

$$(1+x^2)y^{(n+2)}(x)+(2n+1)xy^{(n+1)}(x)+(n^2-\tfrac{1}{4})y^{(n)}(x)=0.$$

Given that $y_0(x)$ is the particular solution such that $y_0(0)=0$ and $y_0'(0)=\frac{1}{2}$, prove that, for all integers $m>1$,

$$y_0^{(2m)}(0)=0, \qquad y_0^{(2m+1)}(0)=-\tfrac{1}{4}(4m-1)(4m-3)y_0^{(2m-1)}(0).$$

Hence prove that the Maclaurin expansion of $y_0(x)$ is

$$\sum_{m=0}^{\infty} \frac{(-1)^m (4m)!\, x^{2m+1}}{2^{4m+1}(2m)!\,(2m+1)!}.$$

314 A function $y(x)$, such that $y(0)=1$, satisfies the equation

$$x(1-x)y''(x)+(2-3x)y'(x)-y(x)=0.$$

Show that

$$(n+1)^{(n)}(0)=n^2 y^{(n-1)}(0) \qquad (n>0).$$

Deduce the Maclaurin series for $y(x)$ and determine its radius of convergence.

315 Prove that

$$(x^2-1)^n = 2^n \sum_{r=0}^{n} (-\tfrac{1}{2})^r \frac{(-n)_r}{r!} (x-1)^{n+r}$$

and deduce *Rodrigues' formula*

$$P_n(x)=\frac{1}{2^n n!}\frac{\mathrm{d}^n}{\mathrm{d}x^n}(x^2-1)^n.$$

Show further that

$$P_n(x)=\frac{(2x)^n(\frac{1}{2})_n}{n!}\,{}_2F_1(\tfrac{1}{2}-\tfrac{1}{2}n,\ -\tfrac{1}{2}n, -\tfrac{1}{2}n; \tfrac{1}{2}-n; x^{-2}).$$

316 The *Laguerre polynomial* $L_n(x)$ is the solution of the initial-value problem

$$xy''(x)+(1-x)y'(x)+ny(x)=0, \qquad y(0)=1$$

in which n is a fixed positive integer. Show that

$$L_n(x)=\sum_{r=0}^{n}\frac{(-n)_r}{r!\,r!}x^r.$$

Write down the nth derivative of $x^n \mathrm{e}^{-x}$ and deduce that

$$L_n(x)=\frac{\mathrm{e}^x}{n!}\frac{\mathrm{d}^n}{\mathrm{d}x^n}(x^n \mathrm{e}^{-x}).$$

317 If m and n are positive integers $(m<n)$

$$L_n^m(x)=\frac{\mathrm{d}m}{\mathrm{d}x^m}L_n(x)$$

is called the *associated Laguerre polynomial.* Show that $L_n^m(x)$ is a solution of the differential equation

$$xy'' + (m+1-x)y' + ny = 0.$$

Show also that the Laguerre function

$$R_{nl}(x) = e^{-1/2x} x^l L_{n+1}^{2l+1}(x)$$

is a solution of the ordinary differential equation

$$\frac{d^2R}{dx^2} + \frac{2}{x}\frac{dR}{dx} - \left\{\frac{1}{4} - \frac{n}{x} + \frac{l(l+1)}{x^2}\right\} R = 0.$$

318 The Hermite polynomial of even degree $2n$, denoted by $H_{2n}(x)$ is the solution of the initial-value problem

$$y'' - 2xy' + 4ny = 0, \qquad y(0) = 1, \quad y'(0) = 0$$

in which n is a fixed positive integer.

Express $y^{(2r)}(0)$ in terms of $y^{(2r-2)}(0)$ and deduce an expression for $H_{2n}(x)$.

If $H_{2n+1}(x)$ is the solution of the initial-value problem

$$y'' - 2xy' + (4n+2)y = 0, \qquad y(0) = 0, \quad y'(0) = 1$$

find the expression for $H_{2n+1}(x)$.

319 Show that, if n is a positive integer, the equation

$$\frac{d^2\psi}{dx^2} + (2n+1-x^2)\psi = 0$$

has solution $\psi = e^{-(1/2)x^2} H_n(x)$.

320 Show that the general solution of Legendre's associated equation

$$(1-x^2)y''(x) - 2xy'(x) + \left\{n(n+1) - \frac{m^2}{1-x^2}\right\} y(x) = 0$$

is given by the equation $y(x) = c_1 P_n^m(x) + c_2 Q_n^m(x)$, where

$$P_n^m(x) = (x^2-1)^{1/2m} \frac{d^m}{dx^m} P_n(x),$$

$$Q_n^m(x) = (x^2-1)^{1/2m} \frac{d^m}{dx^m} Q_n(x).$$

321 Show that Laguerre's initial-value problem

$$t\ddot{y} + (1-t)\dot{y} + ny = 0, \qquad y(0) = 1$$

in which n is a positive integer, has a solution

$$L_n(t) = \mathscr{L}^{-1}[p^{-n-1}(p-1)^n; t].$$

By showing that

$$\mathscr{L}\left[\frac{e^{-xt/(1-x)}}{1-x}; t \to p\right] = \frac{1}{x+(1-x)p}$$

prove that for sufficiently small $|x|$

$$\frac{e^{-xt/(1-x)}}{1-x} = \sum_{n=0}^{\infty} L_n(t)x^n.$$

322 Solve the equation

$$y'' + 2xy' + x^2 y = 0$$

by first reducing it to normal form, and deduce the form of the Green's function for the initial-value problem at $x = a$.

Find the general solution of each of the equations 323–336:

323 Solve the differential equation

$$y\frac{d^2y}{dx^2} + \left(\frac{dy}{dx}\right)^2 = 1$$

given that $y = 1$, $dy/dx = 0$ when $x = 0$.

324 Show that

$$y = \frac{1}{1+x}$$

is a solution of the differential equation

$$x(1+x)y'' - 2y' - 2y = 0,$$

and hence find the general solution of

$$x(1+x)y'' - 2y' - 2y = 1 + x.$$

Verify that $y = x^3$ is a solution of

$$x(1+x)y'' - 2y' - 2y = 4x^3,$$

and deduce the general solution of this equation from that of the previous equation.

325 Assuming that the functions $Q(x)$ and $R(x)$ have continuous derivatives, find a necessary and sufficient condition on them for

$$\frac{d^2y}{dx^2} + Q(x)\frac{dy}{dx} + R(x)y$$

to be expressible in the form

$$\left(\frac{\mathrm{d}}{\mathrm{d}x}+a\right)\left[\frac{\mathrm{d}y}{\mathrm{d}x}+M(x)y\right]$$

where a is a given constant, and $M(x)$ is a function to be determined.
Show that the expression

$$\frac{\mathrm{d}^2y}{\mathrm{d}x^2}+\frac{x+2}{x+1}\frac{\mathrm{d}y}{\mathrm{d}x}+\frac{xy}{(x+1)^2}$$

satisfies these conditions for a certain value of a.
Hence, or otherwise, find the general solution of the equation

$$\frac{\mathrm{d}^2y}{\mathrm{d}x^2}+\frac{x+2}{x+1}\frac{\mathrm{d}y}{\mathrm{d}x}+\frac{xy}{(x+1)^2}=0.$$

326 Find the general solution of the equation

$$\mathbf{L}y=2x+1,$$

where the operator $\mathbf{L}$ is defined by

$$\mathbf{L}=x^2D^2+x(4+x)D+2(1+x)$$

by finding a factor of $\mathbf{L}$ of the form $xd+a$, where a is a constant.
Find the solution which remains finite at $x=0$.
Find two linearly independent solutions of each of the equations 327–336:

327 $2x^2(1+x^2)y''+xy'-12x^2y=0$
328 $(1-x^2)y''-2xy'+6y=0$
329 $xy''+y'-4xy=0$
330 $(1-x^3)y''+6xy=0$
331 $x^2y''+x(1-x)y'-y=0$
332 $(1-x^2)y''-3xy'-y=0$
333 $x^2(1+x)y''-x(1+2x)y'+(1+2x)y=0$
334 $xy''+\frac{1}{2}y'-y=0$
335 $x(1-x)y''+(1-\lambda-2x)y'+\lambda(\lambda-1)y=0$
(λ not zero nor an integer)
336 $x^2y''=\{x^2+p(p+1)\}y$ ($2p$ not an integer)

337 Obtain in the form of infinite series two corresponding independent solutions of the differential equation

$$2x^2y''-(2x+1)xy'+y=0.$$

Show that one of these solutions can be expressed as $\sqrt{x}\,\mathrm{e}^x$, and the other as

$$x\sum_{s=0}^{\infty}\frac{(2x)^s}{1\cdot 3\cdot 5\ldots(2s+1)}.$$

338 The differential equation

$$x^2y''+2(x+1)y'=n(n+1)y$$

has solution

$$y=f(x)\equiv 1+\sum_{r=1}^{n}a_rx^r$$

Find a formula for a_r.

By means of the substitution $y=\mathrm{e}^{2/x}z$ show that a second solution of the differential equation is $y=\mathrm{e}^{2/x}f(-x)$.

339 Show that the differential equation

$$xy''+2y'+xy=0$$

has a solution

$$1+\sum_{k=1}^{\infty}c_k\frac{x^k}{k!}$$

where

$$(k+3)c_{k+2}=-(k+1)c_k.$$

Hence derive the solution $y_1(x)=(\sin x)/x$.

340 Find a series solution of the differential equation

$$xy''+(\gamma-x)y'-\alpha y=0$$

such that $y=1$ when $x=0$. Denoting this solution by ${}_1F_1(\alpha;\gamma;x)$, show that second solution is $x^{1-\gamma}{}_1F_1(\alpha-\gamma+1;2-\gamma;x)$.

Transform the differential equation by the substitution $y=\mathrm{e}^x v$ and show that $\mathrm{e}^x\,{}_1F_1(\alpha;\gamma;c)={}_1F_1(\gamma-\alpha;\gamma;-x)$.

341 Prove that any solution of the equation

$$xy''+y=0$$

is expressible in the form

$$y=x^{1/2}\{AJ_1(2x^{1/2})+BY_1(2x^{1/2})\}.$$

342 Show that the general solution of the differential equation

$$y''+4x^2y=0$$

is

$$y=x^{1/2}\{AJ_{1/4}(x^2)+BJ_{-1/4}(x^2)\}.$$

343 Prove that the solution of the differential equation

$$y''+9x^4y=0$$

is

$$y=x^{1/2}\{AJ_{1/6}(x^3)+BJ_{-1/6}(x^3)\}.$$

344 Show that any solution of the equation

$$y''+e^{-2x}\,y=0,$$

which remains finite as $x\to\infty$ is of the form

$$y=cJ_0(e^{-x}),$$

where c is an arbitrary constant.

Using the Laplace transform find the general solutions of:

345 $\dot{x}+y=\sin t,\quad \dot{y}-x=\cos 2t$

346 $\dot{x}=4x-2y+e^t,\quad \dot{y}=6x-3y+e^{-t}$

347 $5\dot{x}+3\dot{y}-11x-7y=e^t,\quad 3\dot{x}+2\dot{y}-7x-5y=e^{2t}$

348 $7\dot{x}+\dot{y}+2x=E,\quad \dot{x}+3\dot{y}+y=0\quad (x=y=0 \text{ when } t=0)$

(a) when $E=30$, (b) when $E=29\sin t$

349 $5\ddot{x}+\dot{y}+2x=4\cos t,\quad 3\dot{x}+y=8t\cos t$

$(x=1,\ \dot{x}=0 \text{ when } t=0)$

350 $\ddot{x}+3\dot{y}-4x+6y=10\cos t,\quad \dot{x}+\ddot{y}-2x+4y=0$

$(x=y=0,\ \dot{x}=2 \text{ when } t=0)$

351 $\ddot{x}-\dot{y}-2x=\sin t,\quad \ddot{y}-\dot{x}-2y=\cos t$

$(x=\dot{x}=y=\dot{y} \text{ when } t=0)$

352 $\ddot{x}-x+5\dot{y}=t,\quad \ddot{y}-4y-2\dot{x}=-2$

$(x=\dot{x}=y=\dot{y}=0 \text{ when } t=0)$

353 Find the solution of the simultaneous differential equations

$$\dot{x}+y+z=t+2$$

$$x+\dot{y}+z=t^2+t$$

$$x+y+\dot{z}=-t^2+4t+3$$

satisfying the initial conditions $x=y=z=0$ when $t=0$.

354 The motion of a particle is governed by the pair of equations

$$\ddot{x}-3x=y,\quad \ddot{y}+3y=7x,$$

Given that it starts from rest at (0, 1), find its coordinates at any subsequent time. Show that as $t \to \infty$ it moves along the straight line $x = y$.

355 Under certain conditions the currents x and y, respectively, in the primary and secondary windings of a certain transformer, satisfy the equations

$$L\dot{x} + Rx + \tfrac{1}{2}L\dot{y} = E,$$

$$L\dot{y} + Ry + \tfrac{1}{2}L\dot{x} = 0,$$

where L, R and E are constants.

Solve the equations for x and y given that both are zero when $t = 0$. Show that, for small values of t, x is approximately $4Et/3L$ and that, for large values of t, x is approximately E/R.

Find the general solution of the matrix equation

$$\frac{\mathrm{d}\mathbf{A}}{\mathrm{d}t} = \mathbf{A}$$

when $\mathbf{A}$ is given by the expressions 356–376:

356 $$\mathbf{A} = \begin{bmatrix} 3 & -4 \\ 4 & -5 \end{bmatrix}$$

357 $$\mathbf{A} = \begin{bmatrix} 3 & 2 \\ -2 & 3 \end{bmatrix}$$

358 $$\mathbf{A} = \begin{bmatrix} 1 & 2 \\ -1 & 4 \end{bmatrix}$$

359 $$\mathbf{A} = \begin{bmatrix} 3 & 3 \\ 2 & 2 \end{bmatrix}$$

360 $$\mathbf{A} = \begin{bmatrix} 3 & 1 \\ -1 & 1 \end{bmatrix}$$

361 $$\mathbf{A} = \begin{bmatrix} 1 & 1 & 4 \\ 2 & 0 & -4 \\ -1 & 1 & 5 \end{bmatrix}$$

362 $$A = \begin{bmatrix} 3 & -1 & 2 \\ 2 & 0 & -2 \\ 2 & -1 & -1 \end{bmatrix}$$

363 $$\mathbf{A}=\begin{bmatrix}1 & c & b\\ 0 & 1 & a\\ 0 & 0 & 1\end{bmatrix}$$

364 $$\mathbf{A}=\begin{bmatrix}2 & 2 & 0\\ 1 & 2 & 1\\ 1 & 2 & 1\end{bmatrix}$$

365 $$\mathbf{A}=\begin{bmatrix}2 & -1 & 0\\ 2 & 1 & 1\\ -2 & 2 & 1\end{bmatrix}$$

366 $$\mathbf{A}=\begin{bmatrix}8 & 7 & 7\\ -5 & -6 & -9\\ 5 & 7 & 10\end{bmatrix}$$

367 $$\mathbf{A}=\begin{bmatrix}3 & -1 & -1\\ -2 & 3 & 2\\ 4 & -1 & -2\end{bmatrix}$$

368 $$\mathbf{A}=\begin{bmatrix}2 & -1 & 0\\ -1 & 2 & -1\\ 0 & -1 & 2\end{bmatrix}$$

369 $$\mathbf{A}=\begin{bmatrix}1 & 2 & 2\\ 0 & 2 & 1\\ -1 & 2 & 2\end{bmatrix}$$

370 $$\mathbf{A}=\begin{bmatrix}-3 & 6 & 2\\ -4 & 7 & 2\\ 0 & 0 & 1\end{bmatrix}$$

371 $$\mathbf{A}=\begin{bmatrix}2 & 1 & 0 & 1\\ 1 & 3 & -1 & 3\\ 0 & 1 & 2 & 1\\ 1 & -1 & -1 & -1\end{bmatrix}$$

372 $$\mathbf{A}=\begin{bmatrix}1 & 0 & 1 & 1\\ 0 & 1 & -1 & 1\\ 1 & -1 & 0 & 0\\ 1 & 1 & 0 & 0\end{bmatrix}$$

373

$$\mathbf{A}=\begin{bmatrix} 0 & 0 & 2 & 0 \\ 1 & 0 & 1 & 0 \\ 0 & 1 & -2 & 0 \\ 0 & 0 & 0 & 1 \end{bmatrix}$$

374

$$\mathbf{A}=\begin{bmatrix} 10 & -9 & -0 & 0 \\ 4 & -2 & 0 & 0 \\ 0 & 0 & -2 & -7 \\ 0 & 0 & 1 & 2 \end{bmatrix}$$

375

$$\mathbf{A}=\begin{bmatrix} 1 & 1 & 1 & 1 \\ 1 & 1 & -1 & -1 \\ 1 & -1 & 1 & -1 \\ 1 & -1 & -1 & 1 \end{bmatrix}$$

Solutions

1. $y=(x-1)/(cx+1)$
2. $y=(1-cx^2)/(1+cx^2)$
3. $x^3+3y+c+6\log|1-y|=0$
4. $|x\cos y|=c\,e^{-x^2}$
5. $y^2+1=cx/(x+1)$
6. $\sqrt{(x^2-1)}+\sqrt{(y^2-1)}=c$
7. $y=\{x+c\sqrt{(1-x^2)}\}/\{\sqrt{(1-x^2)}-cx\}$
8. $y=(\cos^2 x-c)/(\cos^2 x+c)$
9. $y=cx(1+x)^{-3/2}(1-x)^{-1/2}$
10. $x^2+y^2=\log\{cx^2(y^2+1)\}$
11. $y+b=c\sqrt{(x^2+a^2)}\{x+\sqrt{(x^2+a^2)}\}$
12. $y+a\log(y^2+a^2)=x-x^{-1}+c$
13. $x^2=2y^2\log(cy)$
14. $x+y=c(x^2+y^2)$
15. $y(y+3x)^5=cx^3$
16. $cy=x^4+6x^2y^2$
17. $x^2=c(c-2y)$
18. $y=c\,e^{x^3/3ay^3}$
19. $(y-x)^{1-a}=cxy$
20. $(y-x)^{2+a}=c(x^2+xy+y^2)^{1-a}$
21. $(y-x)^2=cx\,e^{-y/x}$
22. $x^{-b}y^a=c\exp(x/y-y/x)$
23. $c(y-x)=\exp\{2x(2y-x)/(y-x)^2\}$
24. $\exp\tan(y/x)=c/x$
25. $(x+y+2)^2(y-x-1)=c$
26. $(x+y-1)^3=c(x-y+1)^2$
27. $2\log(x^2+y^2)+\tan^{-1}(y/x)=c$
28. $(x-2y+11)^2=c(2x+y+2)$
29. $(x+y+3)^3(x-3y+11)^4=c$
30. $(x+2y-10)^5(x-y+11)^3=c$
31. $(x-2y)^2+6x-10y=c$
32. $(a+1)(y-x)+2b\log|(a+1)(y+ax)+(a-1)b|=c$
33. $x^3(x-y+1)=c(y-4x-1)$

34. $\log(y+a)+2\tan^{-1}\{(y+a)/(x+b)\}=c$
35. $(x-y)^2+3x+y+\frac{1}{2}\log(2x-2y+1)=c$
36. $x-y+\log(x^2+2xy+y^2+2x+2y+1)=c$
37. $ax^2+2hxy+by^2=c$
38. $x^3+y^3+2x^2y=c$
39. $(x+y)(ax^2+2hxy+by^2)=c$
40. $2x^3y^2-4x^2y^3+5x^2y+3xy=c$
41. $x^2\log|y|=c$
42. $y\cos x-x\sin y=c$
43. $y+b=c(x+a)^2$
44. $y=(cx-1)/(x-c)$
45. $y=1-x+c\,e^{-x}$
46. $\sqrt{x}(x^3+7y^4)=c$
47. $x^2y^2=\log(cy^2)$
48. $\sin^{-1}x+y\sqrt{(1-x^2)}=c$
49. $xy+\log|y|=c$
50. $\cos y=c(x+y)$
51. $x^3+3xy+y^3=cx^2$
52. $(x-y)\exp\{\frac{1}{2}(x^2+y^2)\}=c$
53. $(x+y)y\sqrt{(x^2+a^2)}=c$
54. $2x^3y^5+x^{-2}-y^{-2}=c$
55. $(x+y)\,e^y=c(y+1)$
56. $(x^3+y)(x+y)^4=c$
57. $y=cx+\frac{1}{2}x^3-1$
58. $y=\frac{1}{2}x(\log|x|-\frac{1}{2})+c/x$
59. $y=x^4+c(x+1)$
60. $y=x+c\sqrt{(x^2+a^2)}$
61. $y=(1-x^2)^{-1/2}(\sin^{-1}x+c)$
62. $y=\left(\dfrac{1+x}{1-x}\right)^{1/2}(\sin^{-1}x+c)+1+x$
63. $y=x^2-2+c(x^2+1)^{-1/2}$
64. $y=\sqrt{(x+1)}+\dfrac{c}{\sqrt{x}}$
65. $y=x+\dfrac{c-\log\{x+\sqrt{(x^2+1)}\}}{x+\sqrt{(x^2+1)}}$
66. $y=\dfrac{1}{2}+\dfrac{2+\log\{\sqrt{x}+\sqrt{(x-1)}\}}{2\sqrt{\{x(x-1)\}}}$

67. $y=1+\frac{1}{2}c+\dfrac{c+\log|x|}{x}$

68. $y=x-\frac{3}{2}+c(x^2+x+1)^{-1/2}$
69. $y=\sqrt{(x^2-1)}[c+\log\{x+\sqrt{(x^2-1)}\}]-x$
70. $y=cx\sqrt{(1-x^2)}-x+x^3$
71. $(1+x^2)y=c+\log|\sec x|$

72. $y=\dfrac{c}{\sqrt{(1-x)}}+\left(\dfrac{1+x}{1-x}\right)^{1/2}$

73. $y=\dfrac{c+x^2}{\sqrt{(1-x)}}$

74. $y=\dfrac{x-1}{x^2+1}+\dfrac{c}{\sqrt{(x^2+1)}}\exp(-\tan^{-1}x)$

75. $x^2y=c-x\cos x+\sin x$
76 $y=(x+c)\cos x$
77. $y=c\cos x-2\cos^2 x$
78. $y=\sec^2 x+c\operatorname{cosec} 2x$
79. $y=\sin x-1+c\exp(-\sin x)$
80. $y=(e^x+c)\sin x$
81. $y=c\,e^x-x-1-\frac{1}{2}(\sin x+\cos x)$
82. $y=(2x+c)\sec 2x+\tan 2x$
83. $y=\frac{1}{2}x(x-1)+\frac{1}{4}+2\cosh x-\sinh x+ce^{-2x}$
84. $y=\sin x+\frac{1}{2}\sec x+c\cos x$
85. $y=(c+e^{3x}+3e^x)\operatorname{sech} x$
86. $y=c\tan x-\sec^2 x$
87. $y=\sec^3 x(c+\log|\sin x|)+\frac{1}{2}\sec x$
88. $y=\cot x\{c+\cos 2x+2\log|\sin x|\}$
89. $y\sin x=c-\frac{1}{8}\cos 4x-\frac{1}{4}\cos 2x+\frac{1}{2}\log|\tan 2x|$
90. $y\cos x=c+\frac{1}{2}\cos x-\frac{1}{6}\cos 3x+\frac{1}{2}\log|\tan(\frac{1}{2}x+\frac{1}{4}\pi)|$
91. $y=\sin^{-1}x+\sqrt{(1-x^2)}\{c+\frac{1}{2}\log(1-x^2)\}+\sin^{-1}x$

92. $y=c\left(1+\dfrac{1}{x}\right)e^x-(x+1)$

93. $y=(cx^2+1)^2$
94. $y=(cx+x^3+1)^{1/2}$
95. $y=x(c-x^2)^{-1}$
96. $y=\{1+c(1+x^2)^{1/2}\}^{-1}$
97. $y=x^2(c-x^2)^{-1/2}$

98. $y=(cx+\log|x|+1)^{-1}$
99. $y=(2x+cx^2)^{-1/2}$
100. $y=(1+c\sqrt{x})^2/x^2$
101. $(x^3+c)y^3+3x^2=0$
102. $y=(c\sec^2 x+2\tan x\sec x)^{-1/2}$.
103. $y=(c\,e^x+x+1)^{-2}$
104. $y=\left(\dfrac{c-x^2+x^3}{1+x}\right)^{1/2}$
105. $y=1+(x+c\,e^x)^{-1}$
106. $x^2y=x+1+(c\,e^{-2x}-\tfrac{1}{2})^{-1}$
107. $y=x+(cx+ax^{m-2})^{-1}$
108. $y=x+(c\,e^{-x}+x-1)^{-1}$
109 $y=x+a\{c(x^2+a^2)^{1/2}+2x\}^{-1}$
110. $y=\dfrac{x(cx+\sin x)}{cx-\sin x}$
111. $x\sin(x+y)=c$
112 $y=x\tan(x+c)$
113. $y^2=c(1-x^2)-2x-(1-x^2)\log\left|\dfrac{1-x}{1+x}\right|$
114. $y^2=\dfrac{c-x^2+x^3}{1+x}$
115. $(x+y)^2+2=c(x-y)^2$
116. $(x^2+y^2)^2+2a(x^2-y^2)=c$
117. $1-a/r^2=c\,e^{2a\theta}$
118. $a/r^2+\log r^2=c+\cos 2\theta$
119. $x^2=y^2(c-y^2)$
120. $y=x^2\sin(x+c)$
121. $y=\dfrac{x^3+cx}{cx^2+1}$
122. Equation becomes $x^m z'=2z$: (i) $y=x\coth(c-x)$;
(ii) $y=x(1+cx^2)/(1-cx^2)$; (iii) $y=x\coth(x^{-1}+c)$
123. $xy=c$
124. $x^2/c^2+y^2/2c^2=1$
125. $\pm x=a\log|y|-a\log\{a+\sqrt{(a^2-y^2)}\}+\sqrt{(a^2-y^2)}$
126. $y=cx^2+2a^2/3x$

127. $x+c=a\log\{y+\sqrt{(y^2-a^2)}\}$
128. $x^{2/3}+y^{2/3}=c^{2/3}$
129. $y^2=4a(x+a)+c\,e^{x/a}$
130. $y=cx+ax\log|x|-b$
131. $(a-y)(2a+y)^2=9a(c\pm x)^2$
132. $ay=c+\frac{1}{2}x(4x^2-a^2)-\frac{1}{4}a^2\{2x+\sqrt{(4x^2-a^2)}\}$
133. $r=c\,e^{\pm k\theta}$
134. $x^2+y^2=cy$ *or* $y=c$
135. $r=a\sin(\theta+c)$
136. $r=c+a\log|\tan\frac{1}{2}\theta|$
137. $1+2a/r=(\theta+c)$
138. $r^m=c\sin(m\theta)$
139. $y(3x^2-y^2)=c$
140. $2x^2+y^2=c^2$
141. $x^2+y^2-2a\log|x|=c$
142. $(x^2+y^2)^2+2a^2(y^2-x^2)=c^4$
143. $y(3x^2+y^2)=\beta(x^2+y^2)^3$
144. $(1-a/x^2)^b=c(1-b/y^2)^a$
145. $x^2+y^2-4x+4\log|x+1|=c$
146. $2y^2+x^2(\log x^2-1)=c$
147. $x+a+1=c\exp(\frac{1}{2}y^2+x)$
148. $y+c=\pm c\cosh x$
150. $3a(y^2-x^2)+2x^3=c^3$
151. $x^2+y^2=2a^2\log(cx)$
152. $a^2\log x^2+b^2\log y^2=c+x^2+y^2$
153. $\beta\pm\frac{1}{2}\theta=\left(\dfrac{2r}{l}-1\right)^{1/2}-\tan^{-1}\left(\dfrac{2r}{l}-1\right)^{1/2}$
154. $r-\dfrac{1}{r}=\beta-\theta$
155. $(r^2-a^2)\cos\theta=\beta r$
156. $r^2(\sin\theta-\frac{1}{2}\sin 2\theta)=c$
157 $r=\beta\sin\frac{1}{2}\theta(\sin\theta)^{1/2}$
158 $y^2-2mxy-x^2=c$
159 $x^2+y^2=c(my-x)$
160. $r^p=c\sin(p\theta+\omega)$
161. $(m\neq\frac{1}{2})$: $\quad y=c|x|^{1/m}-\dfrac{x^2}{2m-1}$; $\qquad (m=\frac{1}{2})$: $\quad y=cx^2-x^2\log x^2$

If $m>0$ every curve passes through the origin which is a *node*;

while if $m<0$ no curve passes through the origin and $y \to \pm\infty$ as $x \to 0$ so the origin is a col.

162. $y^2 = ce^{2x} + (x+\frac{1}{2})$; the origin is a *limiting point*
163. $x^2 = c^2 - 2cy$
164 $c^2 e^{2y} - 2c\, e^y + \cos^2 x = 0$
165. $144(2y - x - x^2 - c)^2 = (1+8x)^3$
166. $y^2 - 2c\sqrt{x}\cos(\frac{1}{2}\sqrt{3x}) + c^2 x = 0$
167 $27y^2 + 4x^3 = 0$
168. $(x+y)^2 - 2(x-y) + 1 = 0$
169. $xy + a = 0$
170. $y^2 = x^2 - 1$
171. $y = \frac{1}{3}x^3 - x$
172. $ax^2 + 2hxy + by^2 = 1$
173. $x^{2/3} + y^{2/3} = a^{2/3}$
174. $y = e^x$
175. $y = \sin x$
176. $y = \cosh x$
177. $y^6 = 12x$
178. General solution $y = c\, e^x + 1/c$; singular solution $y^2 = 4\, e^x$
179. $(x-c)^2 + y^2 = f(c)$
180. $27(x^2+y^2)^2 - 4x(x^2+9y^2) + 4y^2 = 0$
181. $x^2 - 2C^2xy + C^3 = 0$
182. $4(x^2+y)(y^2-cx) = (xy+c)^2$
183. General solution $y = a\, e^{(x+c)/a} - c$; singular solution $y = x + a$
184. General solution $4a(y+c) = (x+c)^2$; singular solution $y = x - a$
185. General solution $(3cy+1)^2 = 4c^3x^3$; singular solution $x + y = 0$
186. $(2x+c)^3 + (2x+c)^2 - 18(2x+c)y = 27y^2 + 16y$
187. $y^2 = c^2 - 2cx$
188. The general solution has parametric equations $x = (\log p^2 - 2p + c)(1-p)^{-2}$, $y = \{(\log p^2 + c - 4)p^2 + 2p\}(1-p)^{-2}$; singular solution $y = x + 2$
189. The general solution has parametric equations $x = (\frac{3}{2}p^2 - p^3 + c) \times (1-p)^{-2}$, $y = (cp^2 + p^3 - \frac{1}{4}p^4)(1-p)^{-2}$; singular solution is $y = x + 1$
190. The general solution has parametric equations $x = q^{-1} - q - q^2$, $y = c - \log|q| - q + \frac{1}{3}q^3 + \frac{1}{2}q^4 - \frac{1}{2}q^{-2}$
191. General solution $(x+c)(y-bc) = a^2$; singular solution $y + bx = \pm 2a\sqrt{b}$
192. General solution has parametric form $x = 2\tan t - c\sin t$, $y = \tan^2 t - 2 + c\cos t$

193. General solution is $r = a \operatorname{cosec}\alpha \sin^2(\theta-\alpha)$, where α is an arbitrary constant; singular solution is $r + a\sin\theta = 0$
194. General solution has parametric equations $x = c\log\{(t+1)^2|t-1|\}$, $\quad y = t(t-1)(t+1)^{-2}$
195. General solution has parametric equations $x = c + \frac{4}{3}\log|t-1| - \frac{1}{9}\log|3t+1|$, $y = t(3t+1)^{1/3}(t+1)^{-4/3}$
196. General solution has parametric equations

$$x = c - t + 3\int\frac{dt}{1+t^3}, \qquad y = \frac{3t^2}{1+t^3}$$

197. General solution has parametric equations

$$x = c - t - \log|t+1| + 4\int\frac{dt}{1+t^4}, \qquad y = \frac{t^2(1+t)}{1+t^4}$$

198. General solution $y = c + x + 1/(c-x)$; singular solution $y = 2(x \pm 1)$
199. $a=1$, $m=0$, $n>0$ with general solution $y^2 = 2cx + c^2 + n$ and singular solution $x^2 + y^2 = n$
200. General solution $x^2 + y^2 - 2cx = m^2(a^2 - c^2)$; singular solution $(m^2+1)x^2 + m^2y^2 = m^4a^2$
201. $y = \log|A\,e^x + B\,e^{-x}|$
202. $x = a + A\,e^y + B\,e^{-y}$
203 $y = (Ax+B)^2$
204. $Ay^2 + a = Bx^A$
205. $x = B + y + A\log|y|$
206. $y^2 = A\cos 2ax + B\sin 2ax$
207. $y = (A+x)/(B+x)$
208. $4(c_1y-2)(c_1y+4) = 9c_1^4(x+c_2)^2$
209. $y = c_1x^{n+1} + c_2$
210. $y = (c_1x + c_2x^a)^{-1}$
211. $y = x(c_1 + c_2\log|x|)^2$
212. $y = c_1 + c_2/x + \{(1-x)/x\}\log|1-x|$
213. $y = c_1\left(\dfrac{3x}{4c_1} - 1\right)^{4/3} + c_2$
214. $y = c_1$ *or* $y = c_1(c_2\,e^{x/c_1} - 1)$
215. (i) real roots m, n, $y = (c_1\,e^{-max} + c_2\,e^{-nax})^{-1/a}$; (ii) $n = m$, $y = e^{mx}(c_1 + c_2x)^{-1/a}$; (iii) complex roots $h \pm ki$, $y = c_1\,e^{hx}|\cos ka(x-c_2)|^{-1/a}$
216. $(x-c_2)^2 + (y-c_2)^2 = c_1^2$
217. $\alpha = (1-m)^{-1}$

218. $y=(x^2+c_1x+c_2)^{-1}$
219. $y=(c_1 \cos.x+c_2 \sin x+3)^{-1/2}$
220. $y=c_1 \exp(c_2 e^x)$
221. $y=(c_1 e^{2x}+c_2 e^{-2x})^3$
222. $c_1^2 y^2-\log y^2=4c_1(x+c_2)$
223. $c_1 y^2+(c_1x-c_2)^2=k$
224. $y=a \cosh(c_1+x/a)+c_2$
225. $c_1^2 y^2=1+c_2 e^{ax}$
226. $x=c_1(t+\frac{1}{2}\sin 2t)+c_2,\ y=c_1 \sin^2 t,\ (p=\tan t)$
227. $(1-x^2)y''+2(m-1)xy'-m(m-1)y=0$
228 $y=c_1(x+3)+c_2 e^x(x^2-4x+6)$
229. $y=c_1(1+x^2)+c_2\{x+(1+x^2)\arctan x\}$
230. $y=(6x^2-6x+1)(c_1+c_2 \log|1-x^{-1}|)$
231. $y=(c_1+c_2x^{n+1})/(1+x)$
232. $y=c_1x(1-x)^{-1}+c_2\{x+1+2[x/(1-x)]\log|x|\}$
233. $y=c_1 e^{-x}+c_2 e^{-2x}+\frac{1}{6}e^x+\frac{1}{10}\sin x-\frac{3}{10}\cos x$
234. $y=c_1 e^{-x}+c_2 e^{4x}-\frac{4}{5}\cos 2x-\frac{3}{5}\sin 2x$
235 $y=e^{-x}(c_1 \cos 2x+c_2 \sin 2x-1)+\frac{1}{2}e^x$
236. $y=e^{-3x}(c_1+c_2x+\cosh x)$
237. $y=c_1 e^{2x}+c_2 e^{3x}+e^x(2x^2+6x+7)$
238. $y=e^{-x}(c_1 \cos\sqrt{2}x+c_2 \sin\sqrt{2}x+\cos x)$
239. $y=e^{2x}(c_1+c_2x+\frac{1}{4}x^2-\frac{1}{8}\cos 2x)$
240. $y=c_1 e^{-2x}+e^x(c_2+c_3x+\frac{1}{27}x^2-\frac{1}{27}x^3+\frac{1}{36}x^4)$
241. $y=e^{x \sin\alpha}\{c_1 \cos(x\cos\alpha)+c_2 \sin(x\cos\alpha)\}+\cot\alpha \cos x$
242. $y=e^{x \tan\alpha}\{c_1 \cos x+c_2 \sin x+(x^2-2)\sec^2\alpha$
243. $y=(c_1+c_2x-3x^2)\cos x+(c_3+c_4x-x^3)\sin x$
244. $y=(c_1+c_2x+\frac{1}{4}x^2)e^x+(c_3+c_4x)e^{-x}+c_5 \cos x+c_6 \sin x$
245. $-\frac{1}{10}e^x(\sin 2x+2\cos 2x)$
246. $\frac{1}{8}e^{2x}(\cos x+\sin x)$
247. $e^x(\frac{1}{2}x^2-x)$
248. $2x \sin x-\frac{1}{2}\cos 3x$
249. $x \sin x-x^2 \cos x$
250. $\frac{1}{6}x^3 \cos x$
251. $x^3-6x-12$
252. $e^x(\frac{1}{2}x^2+\frac{1}{3}x^3+\frac{1}{4}x^4)$
253. $-e^x \log|x|$
254. $-e^x \sin x$
255. $-25 e^x \cos x+e^{-x}(3\cos x-4\sin x)$
256. $e^{3x}(2x^2-4x+3)$
257. $y=\cos mx+2m(e^x-1)\sin mx$
258. $y=e^{-x}(x-1)^2$

259. $y=(x+1)\,e^{ax}\sin mx$
260. $y=(x-1)\,e^{x}+(x+1)^2\,e^{-x}$
261. $e^{-x}+\frac{1}{2}e^{x}(x-1)^2$
262. $y=c_1x+c_2\sqrt{(x+\frac{1}{3}x^2)}$
263. $c_1x^2+x\{c_2-\log x^2-(\log|x|)^2\}$
264. $y=c_1/x+c_2/x^2+\frac{1}{2}(1+1/x)^2\log|1+x|-\frac{3}{4}$
265. $y=c_1\cos(\log|x|)+c_2\sin(\log|x|)+\log|x|$
266. $y=c_1x^2+c_2x-x\sin x$
267. $y=\{c_1+\log|x+1|\}(x+1)+c_2\sqrt{(x+1)}-1$
268. $y=x^{\alpha}\{c_1+c_2\log|x|\}+x^{1-\alpha}/(2\alpha-1)^2$
269. $y=\sqrt{x}\{c_1+c_2\log|x|+\frac{1}{8}(\log|x|)^2\}$
270. $y=c_1x^n+c_2x^{n+1}+e^{x}x^n$
271. $y=c_1\,e^{\sin x}+c_2\,e^{-\sin x}$
272. $y=c_1x+c_2\sqrt{(1-x^2)}$
273. $y=c_1(1+2x^2)+c_2x\sqrt{(1+x^2)}$
274. $y=\{c_1(1-x^2)+c_2x\}/(1+x^2)$
275. $t=\int\{p(x)\}^{-1/2}\,dx$
276. $y=c_1\cos\sqrt{x}+c_2\sin\sqrt{x}$
277. $y=c_1\cos t+c_2\sin t,\qquad t=\log|\tan x|$
278. $y=(c_1+c_2x)\operatorname{cosec} x$
279. $y=x^n(c_1\cos mx+c_2\sin mx)$
280. $y=(c_1+c_2\tan x)\sin x$
281. $y=c_1\sin x+c_2(\cos x+x\sin x)-\frac{1}{2}(x+1)\cos x$
282. $x=c_1\,e^{y}+c_2\,e^{-y}+\frac{1}{3}e^{2y}$
283. $q'+2pq=0$, also (a) $q<0$, (b) $q>0$
284. $y=\int e^{xt}\,t^{\alpha-1}(t-1)^{\gamma-\alpha-1}\,dt$. Terminals $(0,1)$, also $(-\infty,0)$ if $x>0$, or $(1,\infty)$ if $x<0$
285. $x\tan x+\log|\cos x|$
286. $\sin 2x\log|\cos x|-x\cos 2x$
287. $\log|\cos x|+\sin x\log|\tan(\frac{1}{4}\pi+\frac{1}{2}x)|-1$
288. $\{4\sin\sqrt{x}\}/x^{3/2}-\{2\cos\sqrt{x}\}/x$
289. $$y=\frac{1}{\omega}\int_0^x e^{-k(x-t)}\cos\omega(x-t)f(t)\,dt$$
290. $$y=c_1\,e^{x}\int_1^x e^{-t}t^{-1}\,dt+c_2\,e^{x}+e^{x}\log x$$
291. $y=c_1(2x+5)+c_2\,e^{2x}-e^{\frac{x}{2}}$

292. $y=\left\{c_1 x^{1-2a}+c_2+\dfrac{x^2}{2(2a+1)}\right\} e^{nx}$

$a=+\frac{1}{2} \qquad y=(\frac{1}{4}x^2+c_1 \log|x|+c_2)\, e^{nx}$

$a=-\frac{1}{2} \qquad y=(\frac{1}{2}x^2 \log|x|+c_1 x^2+c_2)\, e^{nx}$

293. $v'+P(x)v-v^2=Q; \qquad u=P-v$

294. $y=c_1 \operatorname{cosec} x+c_2\, e^{2x}(2-\cot x)-\frac{1}{2} e^x (1-\cot x)$

295. $y(1+\sin x)=c_1\{\sin x+2\log(1-\sin x)\}+c_2$

296.

297. $G(x,t)=(x-t)\, e^{-1/2(x^2-t^2)}$

298. $y(x)=\begin{cases}(b-a)^{-1}\{a^{-1}(1-e^{-ax})-b^{-1}(1-e^{-bx})\} & \text{if } 0<x<1\\ (b-a)^{-1}\{a^{-1}(e^a-1)\, e^{-ax}-b^{-1}(e^b-1)\, e^{-bx} & \text{if } x>1\end{cases}$

299. (a) x^{-2} is an integrating factor;
(b) x^{-1} is a particular solution.
General solution is $y=x^{-1}(c_1+c_2\, e^{-1/2x^2})$

300. $y=c_1(x^2+1)+c_2\{x^2+1)\tan^{-1} x\}$

301. $u(x)=cx^{1/2}\exp\left[-\dfrac{1}{2}\displaystyle\int \dfrac{a(x)}{x}\,dx\right],$

$a(x)=\dfrac{2}{1+x}, \qquad u(x)=\frac{1}{2}x^{-1/2}(x+1)$

303. $(x^2+1)^2 y''-4x(x^2+1)y'+(6x^2-2)y=0;$

$y_1=x^2+1, \quad y_2=x(x^2+1)$

310. $y(x)=\displaystyle\sum_{n=0}^{\infty}\frac{2^{2n}(n!)^2}{(2n)!}x^{2n}$, convergent for $|x|<1$

311. $y^{(2m)}(0)=(-1)^m 2^m(-\frac{1}{2}k)_m(\frac{1}{2}k)_m;$
$y^{(2m+1)}(0)=(-1)^m 2^{2m} k(-\frac{1}{2}k+k)_m(\frac{1}{2}k+\frac{1}{2})_m$

312. Solution is ${}_2F_1(\frac{1}{2}a, -\frac{1}{2}a; \frac{1}{2}; x^2)$; this is a polynomial of degree n if $a=\pm 2n$

314. $y(x)=\displaystyle\sum_{n=0}^{\infty}\frac{x^n}{n+1}$, convergent for $|x|<1$

322. $G(x,t)=e^{-1/2(x^2-t^2)}\sinh(x-t)$

323. $y^2=x^2+1$

324. $(c_1+c_2x^3)/(1+x)-\frac{1}{2}x; \qquad y=\dfrac{c_1+c_2x^3}{1+x}+x^3$

325. $y = c_1 \dfrac{x+2}{x+1} e^{-x} + \dfrac{c_2}{x+1}$

326. $y = \dfrac{x^2 - x + 1 - e^{-x}}{x^2}$

327. $y_1(x) = 1 + 2x^2 + \frac{4}{7}x^4 - \frac{88}{77}x^6 + \frac{16}{385}x^8 - \cdots;$
$y_2(x) = x^{1/2}\{1 + \frac{5}{4}x^2 + \frac{5}{32}x^4 - \frac{5}{128}x^6 + \cdots\}$

328. $y_1(x) = 1 - 3x^2 \qquad y_2(x) = x(1 - \frac{2}{3}x^2 - \frac{1}{5}x^4 - \frac{4}{35}x^5 + \cdots)$

329. $y_1(x) = \sum_{r=0}^{\infty} \dfrac{x^{2r}}{(r!)^2},$

$$y_2(x) = y_1(x) \log x - x^2 - \sum_{r=2}^{\infty} \phi(r) \frac{x^{2r}}{(r!)^2}, \quad \text{where}$$

$$\phi(r) = 1 + \tfrac{1}{2} + \tfrac{1}{3} + \cdots + \frac{1}{r}$$

330. $y_1(x) = 1 - x^3, \qquad y_2(x) = x - \frac{1}{2}x^4 - \frac{1}{14}x^7 - \frac{1}{35}x^{10} - \cdots$

331. $y_1(x) = x\left[1 + \dfrac{x}{3} + \dfrac{x^2}{3 \cdot 4} + \dfrac{x^3}{3 \cdot 4 \cdot 5} + \cdots\right] =$

$$= 2x^{-1} \sum_{r=2}^{\infty} \frac{x^r}{r!} = 2x^{-1}(e^x - 1 - x)$$

$$y_2(x) = x^{-1}\left[1 + x + \frac{x^2}{2!} + \frac{x^3}{3!} + \cdots\right] = x^{-1} e^x$$

332. $y_1(x) = (1 - x^2)^{-1/2}, \qquad y_2(x) = (1 - x^2)^{-1/2} \sin^{-1} x$

333. $y_1(x) = x, \qquad y_2(x) = x^2 + x \log x$

334. $y_1(x) = \sum_{r=0}^{\infty} \dfrac{2^{2r} x^r}{(2r)!} = \cosh(2\sqrt{x}),$

$$y_2(x) = \sum_{r=0} \frac{2^{2r+1} x^{r+1/2}}{(2r+1)!} = \sinh(2\sqrt{x})$$

335. $y_1(x) = (1 - x)^{-\lambda}, \qquad y_2(x) = x^{\lambda}\,{}_2F_1(2\lambda, 1; \lambda + 1; x)$

336. $y_1(x) = x^{p+1} F(p + \frac{3}{2}, \frac{1}{4}x^2); \qquad y_2(x) = x^{-p} F(\frac{1}{2} - p, \frac{1}{4}x^2)$, where

$$F(q, t) = 1 + \frac{1}{q} \cdot \frac{t}{1} + \frac{1}{q(q+1)} \cdot \frac{t^2}{2!} + \cdots + \frac{1}{(q)_r} \cdot \frac{t^r}{r!} + \cdots$$

338. $a_1 = \dfrac{(-n)_r (n+1)_r}{r!} \left(-\dfrac{1}{2}\right)^r, \qquad (r \geqslant 1);\ a_0 = 1$

345. $x = A\cos t + B\sin t - \frac{1}{3}\cos 2t$
$y = A\sin t - B\cos t + \frac{1}{3}\sin 2t$

346. $x = A + (B + 4t)\,e^t - e^{-t}$
$y = 2A + \frac{3}{2}(B - 1 + 4t)\,e^t - \frac{5}{2}e^{-t}$

347. $x = A\,e^{3t} + (B - t)\,e^{2t} - \frac{3}{2}e^t$
$y = -2A\,e^{3t} - (B + 2 - t)\,e^{2t} + 2\,e^t$

348. (a) $x = 15 - 10\,e^{-t/4} - 5\,e^{-2t/5}$, $\quad y = -10\,e^{-t/4} + 10\,e^{-2t/5}$
(b) $x = \frac{1}{17}(\frac{116}{3}e^{-t/4} + \frac{85}{3}e^{-2t/5} + 21\sin t - 67\cos t)$
$y = \frac{1}{17}(\frac{116}{3}e^{-t/4} - \frac{170}{3}e^{-2t/5} - 13\sin t + 18\cos t)$

349. $x = (1 - t^2)\cos t$, $\quad y = 3(1 - t^2)\sin t + 14\,t\cos t$

350. $x = 2\,e^{2t} - 2\cos t - 3t\sin t$, $\quad y = t\cos t + (1 - 2t)\sin t$

351. $x = \frac{1}{5}(\sinh 2t - 2\sin t)$, $\quad y = \frac{1}{5}(\cosh 2t - \cos t)$

352. $x = 5\sin t - 2\sin 2t - t$, $\quad y = \frac{1}{2} - \frac{4}{3}\cos t + \frac{5}{6}\cos 2t$

353. $x = -\frac{2}{3}e^t - \frac{1}{3}e^{-2t} + 1 + 2t$, $\quad y = -\frac{2}{3}e^t - \frac{1}{3}e^{-2t} + 1 - t^2$,
$z = \frac{4}{3}e^t - \frac{1}{3}e^{-2t} - 1 + t + t^2$

354. $x = \frac{1}{8}(\cosh 2t - \cos 2t)$, $\quad y = \frac{1}{8}(\cosh 2t + 7\cos 2t)$

355. $x(t) = (E/R)(1 - \frac{1}{2}e^{-2\alpha t} - \frac{1}{2}e^{-2/3\alpha t})$;
$y(t) = -(E/2R)(e^{-2/3\alpha t} - e^{-2\alpha t})$, $\quad \alpha = R/L$

356.
$$e^{-t}\begin{bmatrix} 1+4t & -4t \\ 4t & 1-4t \end{bmatrix}$$

357.
$$e^{3t}\begin{bmatrix} \cos 2t & \sin 2t \\ -\sin 2t & \cos 2t \end{bmatrix}$$

358.
$$\begin{bmatrix} 2\,e^{2t} - e^{3t} & 2(e^{3t} - e^{2t}) \\ e^{2t} - e^{3t} & 2\,e^{3t} - e^{2t} \end{bmatrix}$$

359.
$$\begin{bmatrix} 2\,e^{4t} - e^t & e^{4t} - e^t \\ 2(e^t - e^{4t}) & 2\,e^t - e^{4t} \end{bmatrix}$$

360.
$$e^{2t}\begin{bmatrix} 1+t & t \\ -t & 1-t \end{bmatrix}$$

361.
$$\begin{bmatrix} 2\,e^{2t} - e^{3t} & -2(e^{2t} - e^t) & -\frac{1}{2}(e^{3t} - e^t \\ 2(e^{2t} - e^t) & 2\,e^t - e^{2t} & \frac{1}{2}(e^{3t} - e^t) \\ \frac{1}{2}(e^{3t} - e^t) & -4(e^{2t} - e^t) & 2\,e^{3t} - e^t \end{bmatrix}$$

362.
$$\begin{bmatrix} 3\,e^t - 2 & -(e^t - 1) & -2(e^t - 1) \\ 2(e^t - 1) & 1 & -2(e^t - 1) \\ 2(e^t - 1) & -(e^t - 1) & -(e^t - 2) \end{bmatrix}$$

363. $$e^t\begin{bmatrix} 1 & ct & bt+\frac{1}{2}act^2 \\ 0 & 1 & at \\ 0 & 0 & 1 \end{bmatrix}$$

364. $$\begin{bmatrix} \frac{1}{4}e^{4t}+e^t-\frac{1}{4} & \frac{1}{2}e^{4t}-\frac{1}{2} & \frac{1}{4}e^{4t}-e^t+\frac{3}{4} \\ \frac{1}{4}e^{4t}-\frac{1}{2}e^t+\frac{1}{4} & \frac{1}{2}e^{4t}+\frac{1}{2} & \frac{1}{4}e^{4t}+\frac{1}{2}e^t-\frac{3}{4} \\ \frac{1}{2}e^{4t}-\frac{1}{2} & \frac{1}{2}e^{4t}-\frac{1}{2} & \frac{1}{4}e^{4t}+\frac{3}{4} \end{bmatrix}$$

365. $$\begin{bmatrix} 2(1+t)e^t-e^{2t} & e^t-e^{2t} & (1+t)e^t-e^{2t} \\ 2t\,e^t & e^t & t\,e^t \\ 2e^{2t}-2e^t-4t\,e^t & 2(e^{2t}-e^t) & 2e^{2t}-(1+2t)e^t \end{bmatrix}$$

366. $$\begin{bmatrix} e^{8t} & e^{8t}-e^t & e^{8t}-e^t \\ e^{3t}-e^{8t} & 2e^t-e^{8t} & 2e^t-e^{3t}-e^{8t} \\ e^{8t}-e^{3t} & e^{8t}-e^t & e^{8t}+e^{3t}-e^t \end{bmatrix}$$

367. $$\begin{bmatrix} \frac{1}{2}e^{3t}+\frac{2}{3}e^{2t}-\frac{1}{6}e^{-t} & -(e^{3t}-e^{2t}) & -(\frac{1}{2}e^{3t}-\frac{1}{3}e^{2t}-\frac{1}{6}e^{-t}) \\ -\frac{1}{2}(e^{3t}-e^{-t}) & e^{3t} & \frac{1}{2}(e^{3t}-e^{-t}) \\ \frac{1}{2}e^{3t}+\frac{2}{3}e^{2t}-\frac{7}{6}e^{-t} & -(e^{3t}-e^{2t}) & -(\frac{1}{2}e^{3t}-\frac{1}{3}e^{2t}-\frac{7}{6}e^{-t}) \end{bmatrix}$$

368. $$e^{2t}\begin{bmatrix} \frac{1}{2}(1+\cosh\sqrt{2}t) & \frac{1}{2}\sqrt{2}\sinh\sqrt{2}t & \frac{1}{2}(\cosh\sqrt{2}t-1) \\ -\frac{1}{2}\sqrt{2}\sinh\sqrt{2}t & \cosh\sqrt{2}t & -\sinh\sqrt{2}t \\ \frac{1}{2}(\cosh\sqrt{2}t-1) & -\sinh\sqrt{2}t & (1+\cosh\sqrt{2}t) \end{bmatrix}$$

369. $$\begin{bmatrix} 2(1-t)e^{2t}-e^t & 2e^t-2(1-2t)e^{2t} & 2t\,e^{2t} \\ (1-t)e^{2t}-e^t & 2e^t-(1-2t)e^{2t} & t\,e^{2t} \\ -(e^{2t}-e^t) & 2(e^{2t}-e^t) & e^{2t} \end{bmatrix}$$

370. $$\begin{bmatrix} -(2e^{3t}-3e^t) & 3(e^{3t}-e^t) & e^{3t}-e^t \\ -2(e^{3t}-e^t) & 3e^{3t}-2e^t & e^{3t}-e^t \\ 0 & 0 & e^t \end{bmatrix}$$

371. $$\begin{bmatrix} (1+t^2)e^{2t} & t\,e^{2t} & t^2e^{2t} & t\,e^{2t} \\ 4-(4-9t)e^{2t} & 3(1-t)e^{2t}-2 & (4-9)e^{2t}-4 & 3(1-t)e^{2t}-3 \\ t^2e^{2t} & t\,e^{2t} & (1-t^2)e^{2t} & t\,e^{2t} \\ (4-7t)e^{2t}-4 & (3t-2)2^t+2 & 2(3t-2)e^{2t}+4 & (3t-2)e^{2t}+2 \end{bmatrix}$$

372. $$\begin{bmatrix} \frac{1}{3}(2e^{2t}+e^{-t}) & 0 & \frac{1}{3}(e^{2t}-e^{-t}) & \frac{1}{3}(e^{2t}-e^{-t}) \\ 0 & \frac{1}{3}(2e^{2t}+e^{-t}) & 0 & \frac{1}{3}(e^{2t}-e^{-t}) \\ \frac{1}{3}(e^{2t}-e^{-t}) & -\frac{1}{3}(e^{2t}-e^{-t}) & \frac{1}{3}(e^{2t}+2e^{-t} & 0 \\ \frac{1}{3}(e^{2t}-e^{-t}) & \frac{1}{3}(e^{2t}-e^{-t}) & 0 & \frac{1}{3}(e^{2t}+2e^{-t}) \end{bmatrix}$$

373. $$\begin{bmatrix} \frac{1}{4}(-t)e^{t}+\frac{5}{4}e^{-t}-\frac{1}{2}e^{-2t} & \frac{1}{4}(1-t)e^{t}-\frac{3}{4}e^{-t}+\frac{1}{2}e^{-2t} & \frac{1}{4}(1-t)e^{t}+\frac{5}{4}e^{-t}-\frac{3}{2}e^{-2t} & 0 \\ \frac{1}{8}(5-3t)e^{t}-\frac{5}{8}e^{-t} & \frac{1}{8}(5-3t)e^{t}+\frac{3}{8}e^{-t} & \frac{1}{8}(5-3t)e^{t}-\frac{5}{8}e^{-t} & 0 \\ \frac{1}{8}e^{t}-\frac{5}{8}e^{-t}+\frac{1}{2}e^{-2t} & \frac{1}{8}e^{t}+\frac{3}{8}e^{-t}-\frac{1}{2}e^{-2t} & \frac{1}{8}e^{t}-\frac{5}{8}e^{-t}+\frac{3}{2}e^{-2t} & 0 \\ 0 & 0 & 0 & e^{t} \end{bmatrix}$$

374. $$\begin{bmatrix} (1+6t)e^{4t} & -4t\,e^{4t} & 0 & 0 \\ 4t\,e^{4t} & (1-2t)e^{4t} & 0 & 0 \\ 0 & 0 & \cos t-\frac{2}{\sqrt{3}}\sin t & -\frac{7}{\sqrt{3}}\sin t \\ 0 & 0 & \frac{1}{\sqrt{3}}\sin t & \cos t+\frac{2}{\sqrt{3}}\sin t \end{bmatrix}$$

375. $$\frac{1}{4}\begin{bmatrix} 3e^{2t}+e^{-2t} & e^{2t}-e^{-2t} & e^{2t}-e^{-2t} & e^{2t}-e^{-2t} \\ e^{2t}-e^{-2t} & 3e^{2t}+e^{-2t} & e^{-2t}-e^{2t} & e^{2t}-e^{2t} \\ e^{2t}-e^{-2t} & e^{-2t}-e^{2t} & 3e^{2t}+e^{-2t} & e^{-2t}-e^{2t} \\ e^{-2t}-e^{2t} & e^{-2t}-e^{2t} & e^{-2t}-e^{2t} & 3e^{2t}+e^{-2t} \end{bmatrix}$$

Index